# Foundations of Engineering Mechanics

Morozov · Petrov, Dynamics of Fracture

## Springer

*Berlin*
*Heidelberg*
*New York*
*Barcelona*
*Hong Kong*
*London*
*Milan*
*Paris*
*Singapore*
*Tokyo*

N. Morozov · Y. Petrov

# Dynamics of Fracture

With the assistance of:
V. Stenkin, M. Morozova, P. Stahle, J. Wemmie

With 35 Figures

Springer

*Series Editors:*
V. I. Babitsky

Loughborough University
Department of Mechanical Engineering
LE11 3TU Loughborough, Leicestershire
Great Britain

J. Wittenburg

Universität Karlsruhe (TH)
Institut für Technische Mechanik
Kaiserstraße 12
D-76128 Karlsruhe / Germany

*Authors:*
Nikita Morozov
Yuri Petrov

Mathematics and Mechanics Faculty
St. Petersburg State University
Bibliotechnaja sq. 2
198904 St. Petersburg / Russia

*Translators:*
V. Stenkin, M. Morozova, P. Stahle, J. Wemmie

ISBN 3-540-64274-9 Springer-Verlag Berlin Heidelberg New York

Library of Congress Cataloging-in-Publication Data
Morozov, Nikita Fedorovich, 1932- Dynamics of fracture /
N. Morozov, Y. Petrov; with the assistance of V. Stenkin ...[et al.].
p. cm. - (Foundation of engineering mechanics) Includes bibliographical references.
    ISBN 3-540-64274-9 (hc.: alk. paper)
1. Fracture mechanics-Mathematical models. 2. Brittleness-Mathematical models.
3. Shock (Mechanics)-Mathematical models. I.Petrov, Y. (Yuri), 1958- II. Title. III. Series
TA409.M673 1999    620.1'126-dc21

© Springer-Verlag Berlin Heidelberg 2000
Printed in Germany

The use of general descriptive names, registered names, trademarks, etc. in this publication does not imply, even in the absence of a specific statement, that such names are exempt from the relevant protective laws and regulations and therefore free for general use.

Typesetting: Camera-ready copy from translator
Cover-design: de'blik, Berlin
Printed on acid-free paper    SPIN: 10674291    62 / 3020 hu - 5 4 3 2 1 0

# PREFACE

Fracture mechanics as a scientific discipline has been developed during the last decade. And many of its basic aspects have already been elucidated. A substantial contribution to this development was realised by Russian scholars (N. A. Zlatin, A. G. Ivanov, B. V. Kostrov, E. M. Morozov, L. V. Nikitin, V. S. Nikiphorovski, V. Z. Parton, L. I. Slepjan, V. E. Fortov, G. P. Cherepanov, E. I. Shemyakin and others) and by foreign ones (J. D. Achenbach, K. B. Broberg, J. W. Dally, L. B. Freund, A. Maue, J. F. Kalthoff, W. G. Knauss, T. Kobayashi, G. C. Sih, D. A. Shockey and others). And, to a great extent, the progress in the field is due the achievements of the St. Petersburg – Leningrad Scientific School of Mechanics Continuum by G. V. Kolosov, V. V. Novozhilov and L. M. Kachanov Contributions of this institute include the establishment of the fundamental principles of fracture analysis as a process, occurring at different scale levels of structure.

However, despite such achievements in the development of the science of fracture, many important problems remain. One of the most notable of these is dynamic fracture. This is usually regarded as a rupture of material under a shock-wave loading which takes place in a relatively short time period and corresponds to the working time of the external shock pulse or shorter than it.

Improvement of the existing measuring devices and the creation of new ones made it possible to study the process of fracture under high-speed loading. Subsequently, vigorous growth occurred in both experimental and theoretical aspects of fracture mechanics. But even with this progress, solutions to the principal problems that remain are yielded slowly. This is because of the considerable technical difficulties in experimentation, requiring expensive precision instruments as well as the complexities of mathematical modeling and computation. Much more experimental data needs to be acquired and we are faced with an urgency to systematise the data already in hand.

Thermoviscoelasticoplastic-medium modeling abounds in such defects as microvoids and microcracks. Also, discrepancies in experimental fast-fracture data obtained by traditional approaches have engendered various complicated schemes to compute dynamic effects. The complex numerical methods used and approaches taken to dynamic fracture analysis turn out to be substantively accessible only to their authors. And with linear elasticity mechanics and brittle fracture these schemes are not perfect. Nevertheless, it is still possible to elaborate methods for high-speed fracture analysis based on simple mechanical engineering principles, so the definitive detour from traditional schemes of energetic balance and from power fracture mechanics in favor of complex rheology and fracture microphysics may be premature. The development of such procedures will permit more effectively an explanation of the many peculiarities of high-speed fracture.

In this book we develop a new phenomenological approach to study brittle-medium fracture initiation under shock pulses. This approach is based on defining invariant parameters which are not dependent on the mode and history of a fracture and permit an investigation of miscellaneous effects observed during experiments with high-speed fracture. This approach provides an opportunity to estimate fracture of both 'intact' media and media having macrodefects such as cracks and sharp notches. A qualitative explanation is thus obtained for a number of principally important effects of high-speed dynamic fracture that can not be clarified within the framework of previous approaches. We show that it is possible to apply this new strategy to resolve applied problems of disintegration, erosion, and dynamic strength determination of structural steels, etc. By extending well-known classical principles of Linear Fracture Mechanics, the suggested approach conserves the intrinsic 'industrial' character of the analysis and can be considered as a basis for new testing methods and for certification of dynamic strength characteristics of structural materials.

Specialists can use the methods described in this book to determine critical characteristics of dynamic strength and optimal effective fracture conditions for rigid bodies. This book can also be used as a special educational course for guidance on the deformation of materials and constructions, and fracture dynamics.

We, the authors, take the opportunity to pay tribute to the honor and memory of our teacher, academician V. V. Novozhilov, who approved of the structural–temporal approach described in this book, and to the academician N. S. Solomenko, who gave his support to these investigations.

Numerous invaluable comments on the book were provided by Prof. R. V. Goldstein. It is our agreeable duty to thank him.

We are also grateful to our colleagues and collaborators at the Mathematical and Mechanics Research Institute (NIIMM) and the Mathematics and Mechanics Faculty of the St. Petersburg University. We also thank Dr. A. A. Utkin, who did numerous computations, and T. A. Ephimova for assistance with the computer typesetting of this book.

# CONTENTS

# CHAPTER 1
# PROBLEMS OF DYNAMIC LINEAR FRACTURE MECHANICS

The behavior of materials and constructions under high-speed loading has a number of peculiarities, and its examination is far from complete. This is due to the deficiency of experimental research data on controlled loading in the micro- and nanosecond ranges and by unsuccessful attempts to elucidate the results of these experiments on the basis of traditional static mechanics concepts.

## §1. Problems of Stability and Fracture Under High-Speed Loading

Many experimenters have noticed the fact that rapidly loaded models and constructions can resist, under static conditions, buckling loads which considerably exceed the critical level, but still do not buckle.

Studies on standard rectilinear rod compression ([16, 2, 41]) and others revealed some peculiarities of its behavior in dynamics. It is commonly known that under static conditions, during any exceeding of the Eulerian load for a compressed rod, only the first mode of buckling is stable. It might be supposed that such a state would remain even during dynamic loading. However, M. A. Lavrentiev and A. Y. Ishlinsky [16] have ascertained that for rod compression, calculating the inertial forces acting according to Heaviside's law leads to a different result.

Strength characteristics of materials and constructions under static and dynamic loading also differ. Numerous experiments demonstrate the failure of specimens through fracture under high-intensity pulse loading, when amplitudes of the external effects exceed those forces that would normally cause a fracture under static loading conditions. A thorough investigation of this problem was hindered for a long time by the absence of reliable loading-rate control, and the inability to choose and to measure fracture parameters which uniquely characterise the beginning of the process.

Systematic study of such high-speed fracture peculiarities demands complicated high-precision equipment and has become possible only quite recently. Important research by Russian and other scholars is devoted to this problem (e.g., [8–10, 53, 35, 75–77, 73, 82, 112, 113, 124, 66–68]).

The experimental results affirm that the testing of dynamic strength characteristics on the basis of quasistatic defining parameters is rather problematic.

Traditional parameters of strength and crack-growth resistance are constants for a particular material under static conditions. However these same parameters are very complicated under dynamic loading and depend on physical and geometrical characteristics of the external action. At the same time, attempts to describe the dynamic fracture of materials with a set of functional curves correcting for velocity, temporal and other dependent variables of traditional quasistatics have also been unsuccessful. Such work is complicated not only by extremely serious difficulties in experimental tracing of such curves but also by the special nature of dynamic fracture. Dynamic fracture regularities are characterised by special features which make it impossible to apply directly the concepts used to describe material strength under static conditions.

Experiments have revealed that new approaches are needed for developing models for high-speed fracture analysis. These approaches should reflect the structural and temporal peculiarities of the process. Producing such approaches is one of the highest priorities of deformable rigid-body mechanics.

## §2. Outline of Linear Fracture Mechanics

The modern theory of material strength and fracture begins with the investigations of Galileo, who was the first to associate variations in strength with the existence of defects. Nevertheless, the engineering practices of the nineteenth and the beginning of the twentieth centuries are founded on the hypotheses of Ch. Coulomb, O. Mohr and E. Mariotte, who considered and studied media and constructions as 'defectless', and fracture as an instantaneous action by a local stress $\sigma$ of some critical value.

A new stage of fracture mechanics development is associated with the names of A. Griffith and G. Irwin, who suggested an effective method of equilibrium analysis for a cracked elastic domain.

A. Griffith [72] introduced the specific energy of a fracture $\Gamma$, which represented the energy of new surfaces formed during fracture, and suggested using the energetic balance equation to define critical loads

$$\Delta U + \Delta \Gamma = \Delta A,$$

where $U$ is the deformation energy; $A$ is the work of external forces.

G. Irwin [74], using Sneddon's formulas for stresses near the crack tip, suggested the following criterion of crack propagation in a form which was extremely convenient for applications

$$K_\mathrm{I} \leq K_\mathrm{Ic},$$

where $K_I$ is the stress-intensity factor, and $K_{Ic}$ is the static fracture toughness.

We notice that $K$ is not a local but an integral characteristic, depending
on the energetic state of the whole construction.*

Thus, in the problems of static loading there are two fracture (strength)
criteria: $\sigma \leq \sigma_c$ – 'for defectless' media and $K_I \leq K_{Ic}$ – for cracked domains.
In both cases, we are dealing with certain power characteristics attaining the
given critical value, whereupon, according to the theory, a fracture occurs
instantaneously.

The examined classical approach is reasonably well adjusted to the experimental results for brittle fracture and is now universally recognised as an
instrument of engineering practice (see, e.g., [40, 44]).

As will be confirmed by examples further on, the situation is changed in
the case of dynamic loading.


## §3. Theoretical Contradictions of Fracture Dynamics

Let us study the problem of interaction between a longitudinal pulse stress
wave with a crack $(x \leq 0, y = 0)$ in an unbounded elastic plane $xy$. On
the crack faces we have the boundary conditions $\sigma_{xy} = 0$, $\sigma_y = 0$. Let the
components of the displacement vector $u$, $v$ in the incident wave be expressed
by the relations

$$u = 0,$$

$$v = v_0 \Big[ (c_1 t + y) H(c_1 t + y) - (c_1 t + y - c_1 T) H(c_1 t + y - c_1 T) \Big].$$

Then the stress $\sigma_y$ in the wave has a rectangular temporal profile

$$\sigma_y = P \Big[ H(c_1 t + y) - H(c_1 t + y - c_1 T) \Big],$$

where $P = (\lambda + 2\mu) v_0$, $P$ is the amplitude of the incident stress wave and $T$
is the pulse duration.

Here and further $c_1 = \sqrt{(\lambda + 2\mu)/\rho}$, $c_2 = \sqrt{\mu/\rho}$ are the speeds of longitudinal and transverse waves respectively, $\lambda$, $\mu$ are the Lamé constants and
$H(t)$ is the Heaviside function.

At the temporal value $t = 0$ the interaction between the incident wave and
the crack happens, whereupon near the crack tip $(r = 0)$ a singular stress
field appears, characterised by asymptotic formulas

$$\sigma_{ij} = \frac{K_I(t)}{\sqrt{2\pi r}} g_{ij}(\theta) + o(1), \quad r \to 0.$$

---

*Using the stress-intensity factor to estimate limit efforts, applied to cracked elastic
domains, was already suggested by K. Wieghardt [121] in 1907.

Here $r$, $\theta$ are polar coordinates at the crack tip.

Let $T$ tend to zero, holding the complete power pulse of the external action $U = PT$ constant. Then, as was proved by G. P. Cherepanov [60],

$$K_{\mathrm{I}}(t) = \frac{U\,\Phi(c_1, c_2)}{2\sqrt{t}}. \tag{1.1}$$

According to this formula there is always a temporal value when the current value of the stress-intensity factor exceeds an arbitrarily large value. So, according to the classical approach, fracture occurs during any, including an indefinitely small, concentrated action pulse $U$, but this does not correspond to the real situation.

An important singularity of dynamic fracture is the existence of specific links between the stress intensity and the energy flux entering the moving crack tip.

Let us examine for example a crack of a longitudinal displacement (or so-called antiplane deformation), extending in an infinite elastic medium with a constant speed $v$. For this case the energy flux entering the moving crack tip, $T(v)$, can be described by the expression (see, e.g., [55])

$$T(v) = \frac{1}{2\mu} \frac{K_{\mathrm{III}}^2}{2(1 - v/c_2)}, \tag{1.2}$$

where $K_{\mathrm{III}}$ is the power-intensity factor at longitudinal displacement. So, according to the theory of cracks, the energy flux entering the crack tip increases infinitely with an increase of its speed and $T(v) \to \infty$ when $v \to c_2$.

For classical fracture mechanics this situation is quite paradoxical: for any material of an arbitrary strength a certain velocity always exists when a fracture would be possible. Hence, any 'energetic barrier', established by the classical theory of Griffith–Irwin, can easily be overcome at specific velocities for any given intensity of external loading.

It should be noted that opinions usually given to justify the aforesaid contradiction about the dependence of crack speed on the energy flux entering the crack tip (the speed increases with the flux increase) are incorrect. These are based on the supposed behavior of the solution to the inverse problem, which has not yet been obtained in final form.

We also note that the real 'crack tip' speeds are considerably less than their theoretical values, even for very intense external actions, and this guarantees a high-power energy flux entering the tip. This contradiction is the subject of numerous studies, but a reasonably satisfactory explanation has not yet been found.

Experiments show that dynamic fracture mechanics abounds in multitudinous effects that can not be incorporated in classical ideas. Many of them will be analysed in the next chapters. But we would like to point out here an important circumstance; a whole series of effects characteristic of dynamics can be explained, and even computed, with the help of a special generalisation of the principles of linear fracture mechanics based on the concept of the spatial–temporal structure of a fast-rupture process.

CHAPTER 2

# EXPERIMENTAL METHODS OF DYNAMIC FRACTURE RESEARCH

Among the experimental methods of research on dynamic crack-growth resistance and fast fracture, the dynamic photoelastic and caustic methods are the most effective. These methods have been developed in the last two decades. The most important feature of these methods is the ability to directly track behavior possibilitis for quantitative characteristics of the material stress state during fracture. This is attained by a combination of classical methods of optical image processing with high-speed photography techniques. In this chapter we will examine principles and peculiarities of both methods as they are applied to fast-start and crack-propagation problems in brittle solids.

## §1. The Dynamic Photoelastic Method

The basis of the dynamic photoelastic method is the ability of many vitreous polymers to show photoelastic phenomenoen. The effect is stipulated by the fact that under the influence of mechanical stresses in clear materials an optical anisotropy appears. This leads to the appearance of birefringence; a linearly polarised light wave passing through a tensile plane decomposes into two orthogonally polarised rays, each of which propagates at its own speed.

If after the passage we bring both rays together to a common polarization plane, we get an interference pattern. This pattern can be investigated according to well-known methods. The difference in optical distance (phase difference), by mechano-optical rheological laws, corresponds to the state of plate strain. This permits us to define quantitative characteristics of the in plane deflection mode at each point of the model.

With this type of modeling, vitreous polymers with clearly expressed elastic–brittle and photoelastic properties are usually used. In particular, Homalite-100 and modified epoxy KTE are such materials (see, e.g., [84]). Homalite-100 is a transparent vitreous polymer with birefringence. Owing to its processability, it is intensively used in studies by the photoelastic method. It can be simply obtained as large sheets with optical-quality polished surfaces. An important quality of this material is its ability to preserve its properties under sustained loading.

It was demonstrated by A. B. Clark and R. J. Sanford [65] that this material's optical constant does not depend on the rate of loading. The study of Homalite-100 dynamic behavior revealed ([85]) that this material is suitable for analyses of crack propagation through the photoelastic method. Nowadays it is widely used by American and west European researchers as one of the most brittle birefringent materials available.

Other materials, also often used in photoelasticity experiments, are compounds based on an epoxide resin. The epoxide KTE (see, e.g., [84]) is obtained by polymerisation of the resin Epon 828. The polymerisation is achieved with the help of a vulcanising ingredient, polyxypropylenamine.

Plates, made from the epoxide KTE, have a strong birefringence and are effectively used in dynamic investigations using the photoelastic method. The epoxide compounds in comparison with Homalite-100 are more viscous and at the same time more sensitive to the loading rate.

Principal mechano-optical values of Homalite-100 and epoxide KTE are given in Table 1 [84].

**Table 1**

| Parameter | Homalite-100 | Epoxide KTE |
|---|---|---|
| $c_1$ (m/s) | 2150 | 1970 |
| $c_2$ (m/s) | 1230 | 1130 |
| $E_d$ (GPa) | 4.82 | 3.86 |
| $\mu_d$ (GPa) | 1.84 | 1.47 |
| $\nu_d$ | 0.31 | 0.34 |
| $\rho$ (N s$^2$/m$^4$) | 122 | 117 |
| $K_{Ic}$ (MN/m$^{3/2}$) | 0.45 | 1.18 |
| $C_{\sigma d}$ (MN/mm) | 0.45 | 1.18 |

Here $E_d$, $\mu_d$ are the dynamic Young's modulus and the shear modulus respectively; $\nu_d$ is the Poisson coefficient; $K_{Ic}$ is the static fracture toughness (crack-growth resistance); $C_{\sigma d}$ is the material optical constant under dynamic loading; and $\rho$ is the density.

Elastic constants under dynamic loading are determined by measuring longitudinal $c_1$ and transverse $c_2$ stress-wave velocities. This measuring is carried out by observing, using the photoelastic method, the stress-wave propagation in a half-plane loaded dynamically, for example with a charge of explosive substance.

The material optical constant $C_{\sigma d}$ is a parameter linking the optical characteristic isochromat sequential number $N$ with the main stresses by means of the optical rheological law

$$2\tau = \sigma_1 - \sigma_2 = NC_{\sigma d}/h, \tag{2.1}$$

where $\tau$ is the peak shear stress in the plate; $\sigma_1 - \sigma_2$ is the difference of principal stresses; and $h$ is the thickness of the specimen. Measuring $C_{\sigma d}$ under dynamic loading can be done on the basis of simultaneous measurement results of axial deformation and the number of fringes in the specimen, loaded by an uni-axial impact ([65]).

A stress state in the crack-tip neighborhood is determined by means of temporal scanning of isochromats. The pattern is obtained with the help of high-speed photography. Different systems of high-speed photography are used. Examples of good cameras are the Kranz–Shardin multi-spark camera (33 000–85 0000 exposures / s) and a streak camera of the SPR-1(2) type (from 1 to 2 million exposures / s). Dimensions and forms of isochromats recorded on the photographs reflect fairly accurately the instantaneous value of the stress-intensity factor. Again, the ability to determine the crack-tip position during its propagation allows us to measure its length as a time function. A typical pattern of behavior of isochromats in the crack-tip vicinity is shown in Fig. 2.1.

Fig. 2.1

To determine the instantaneous values of the parameters of the stress-intensity factor $K(t)$, the dimensions and forms of the isochromats are measured and characterised. The characteristics of the stress-field intensity are acquired on the basis of analytical and experimental test results. Analytical calculations of isochromat forms are carried out with the help of the Vestergaard function of stresses

$$Z(z) = \frac{K}{\sqrt{2\pi z}} \left[ 1 + \beta \left( \frac{z}{a} \right) \right] + \alpha,$$

where $z = re^{i\theta}$. The expression $K/\sqrt{2\pi z}$ describes the singular part of the stress field near the tip of a crack with the length $2a$, located in the center of a plane plate. Parameters $\alpha$ and $\beta$ permit us to take more precise account of the boundary influence and applied loads. Furthermore, the domain near the crack tip, where the described isochromat analysis is used, expands a little. The parameters $K$, $\alpha$ and $\beta$ determine the characteristic form of isochromat loop near this domain. Expressions for $\tau$ and for the difference $\sigma_1 - \sigma_2$ from (2.1), calculated according to the Vestergaard function, make it possible to compute the stress-intensity factor from measured geometrical characteristics of isochromats near the crack tip.

## §2. The Method of Caustics

Stress-intensity factors for stationary and expanding cracks can be experimentally determined by schlieren optical methods. P. Manogg [90] has worked out a method of schlieren figures, which later became well known in fracture mechanics as the method of caustics ([117]).

Let us consider the basic principles of the method of caustics (see, e.g., [79, 108]). Let the notched specimen from the transparent material, illuminated by exterior forces, be lightened by a parallel optical beam. The cross-section of the specimen cut by a plane passing through the zone around the crack tip is shown in Fig. 2.3.

An increase of the intensitiy of the stress in the zone closer to the crack tip leads to a decrease in the plate thickness, and changes the material, index of refraction. Hence, at a first approximation, the domain of the crack tip is acting as a divergent lens, deflecting the light from the axis of the beam. This causes the appearance of a schlieren figure, limited by an intense light edge (caustic), that can be observed on a screen beyond the specimen (Fig. 2.3).

The boundary between light and shadow for the given figure is determined by the annular domain around the crack tip, the radius of which depends on the distance between the screen and the specimen. The appearance of one caustic is typical for isotropic materials, and of two caustics for anisotropic ones. For transparent materials this method can be used in transmitted light, and for non-transparent materials in reflected light.

P. Manogg [90] computed the form of schlieren figures for a tensile crack supposing the stress distribution near the crack tip is described by Sneddon's formula. Following his course of reasoning:

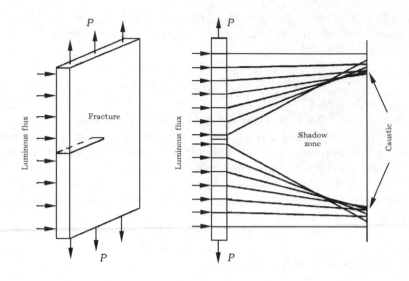

**Fig. 2.2**

Let $x_1$, $x_2$ be the screen coordinates of a ray having been transmitted through a non-deformed plate and $X_1$, $X_2$ be the same coordinates after its deformation. Let us consider the deformed specimen surface in the crack tip ('lens') given by the equation

$$x_3 + f(x_1, x_2) = 0.$$

Regarding the distance from the specimen to the screen $z_0$ as much greater than its thickness ($z_0 \gg f$), we have

$$(X_1, X_2) = (x_1, x_2) - 2z_0 \nabla f.$$

The caustic is the envelope of beams. Its equation is written by assigning the Jacobean coordinate transformation a value of zero,

$$\det \left[ \delta_{\alpha,\beta} - 2z_0 f_{\alpha,\beta} \right] = 0, \quad \alpha, \beta = 1, 2, \tag{2.2}$$

where $\delta_{\alpha,\beta}$ is the Krőneker symbol.

For the opening mode of a tensile crack we have

$$f(r, \theta) = u_3(r, \theta) = -\frac{\nu h}{E\sqrt{2\pi r}} K_{\mathrm{I}} \cos \frac{\theta}{2}, \tag{2.3}$$

where $K_{\mathrm{I}}$ is the stress-intensity factor for a crack of type I; $E$, $\nu$ are constants of elasticity; $h$ is the thickness; and $r$, $\theta$ are polar coordinates at the crack tip.

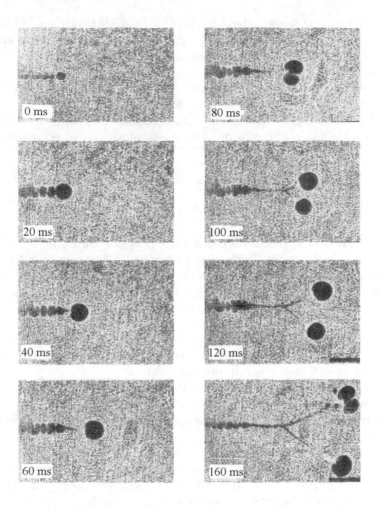

**Fig. 2.3**

A substitution of (2.3) in the transformation (2.2) gives rise to the equation of the caustic, shown on the screen, which turns out to be an epicycloid. The maximum diameter of the caustic is a function of the stress-intensity factor and can be described by the formula

$$K_{\mathrm{I}} = \frac{2\sqrt{2\pi}\,E}{3\lambda^{5/2}\nu h z_0}\,D^{5/2},\qquad(2.4)$$

where $D$ is the caustic diameter; $\lambda$ is a numerical coefficient, characterising the epicycloid form; and $z_0$ is the distance between the specimen and the screen.

Experimental realisation of the method is simple enough. A monochromatic light beam, emitted by a laser, having passed through a system of lenses, falls on a specimen. Reflected (or transmitted) rays are captured on a screen. The specimen is loaded, and the maximal dimensions of the schlieren figure are measured. Then, substituting the parameter values in (2.4), we calculate the stress intensity factor.

During dynamic tests schlieren figures are registered with the help of high-speed photography. The start is initiated by the crack itself.

Materials studied by the caustic method include: polymethylmethacrylate (PMMA), Homalite-100, epoxide, Araldite-B, plexiglas and polycarbonate.

## §3. On an Asymptotic Representation of the Stress Field Near the Crack Tip

A real stress field appearing in a thin plate is always three-dimensional. Experiments on plates from Plexiglas and martensitic steel carried out with the caustic method have shown [86, 106] that the radius of the space-stress-state zone near the macrocrack tip is not less than $0.5h$, where $h$ is the plate thickness. Nevertheless, in practice there are many cases when a brittle-fracture analysis could be carried out with the help of a two-dimensional asymptotic description based on the stress-intensity factor. As was already noted, this can be determined by measuring the stresses in the crack-tip neighborhood according to well-known asymptotic formulas.

Thereby dimensions and shapes of the action zone of a stress state with a two-dimensional asymptotic description acquire great significance. Let us examine this problem in terms of the opening mode of tensile crack behavior in static conditions [48].

This problem has been solved by the photoelastic method using plane specimens made from organic glass. Specimen 1 was made from organic glass with the optical constant $C = -2.04 \times 10^{-2}\,\mathrm{Pa}^{-1}$, and specimen 2 from optically sensitive organic glass with the optical constant $C = 40.5 \times 10^{-2}\,\mathrm{Pa}^{-1}$.

Both specimens were plates with the dimensions $220 \times 68\,\mathrm{mm}$. The thicknesses of the plates 1 and 2 were 3.25 and 4 mm respectively. The crack location and its dimensions are denoted in Fig. 2.4. Specimen 1 crack length $l$ was 25.2 mm, and 26.8 mm for specimen 2. Special holes near the crack faces provided the formation of an ideal crack end from the original notch when stretching forces $p$ were applied.

The specimen was subjected to remote uniaxial stretching in the direction perpendicular to the notch plane by a loading device, UP-8. Each specimen was investigated for two remote tensile stresses $p$ equal to 0.91 and to 2.23 MPa. The stretched specimen was placed in the field of a coordinate synchronous polarimeter, CSP-10. With its help, using the Saint-Armond method, the optical phase difference $\psi$ and the parameter $\varphi$ of optical isocline in monochromatic light with the wavelength $\lambda = 546.1\,\mathrm{nm}$ were measured (see, e.g., [1]). The measurements were taken along the plane $\theta = \mathrm{const}$

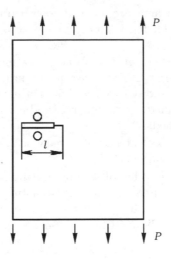

**Fig. 2.4**

($\theta$ is the polar angle). The values of tangential stresses $\tau_{xy}$ and the differences of normal $\sigma_x - \sigma_y$ stresses were calculated at these points according to measured optical values by the photoelastic method

$$\tau_{xy} = \frac{\delta}{2Ch} \sin(2\varphi_{xy}), \quad \sigma_x - \sigma_y = \frac{\delta}{Ch} \cos(2\varphi_{xy}), \tag{2.5}$$

where $\delta$ is the optical propagation difference, determined by the measured phase difference; $C$ is the optical constant obtained from calibrating stretched specimens of the specimen material; and $\varphi_{xy}$ is the angle determining the direction of the largest principal stress $\sigma_1$ relative to the axis $0x$. Moreover, $\varphi_{xy} = \varphi$ for material with $C > 0$, and $\varphi_{xy} = \varphi \mp 90°$ for material with $C < 0$; $h$ being the thickness of the examined model.

It was supposed that the stress-field distribution near the crack tip was described by asymptotic (for $r \to 0$) Sneddon's formulas (see, e.g., [22])

$$\sigma_x = \frac{K_\mathrm{I}}{\sqrt{2\pi r}} \cos\frac{\theta}{2} \left(1 - \sin\frac{\theta}{2} \sin\frac{3\theta}{2}\right) - p + o(1),$$

$$\sigma_y = \frac{K_\mathrm{I}}{\sqrt{2\pi r}} \cos\frac{\theta}{2} \left(1 + \sin\frac{\theta}{2} \sin\frac{3\theta}{2}\right) + o(1), \tag{2.6}$$

$$\sigma_{xy} = \frac{K_\mathrm{I}}{\sqrt{2\pi r}} \cos\frac{\theta}{2} \sin\frac{\theta}{2} \cos\frac{3\theta}{2} + o(1).$$

The relations (2.5) and (2.6) permit a linking between measured optical values and the stress-intensity factor $K_\mathrm{I}$. The correctness of the asymptotic description (2.6) in an area around the crack tip can be confirmed by the stability of values $K_I$, having been calculated according to optical values in such an area.

Point coordinates where optical values were measured have been determined by the polarimeter CSP-10 with an accuracy of 0.02 mm. The accuracy of values $\psi$ and $\varphi$ did not exceed 0.5°. During the experiment the error of stress calculations for $\sigma_x - \sigma_y$ and $\tau_{xy}$, and consequently the error of the value $K_I$, was established.

Measuements of the optical values were carried out along $\theta$ equal to ±135, ±120, ±90, ±75, ±45 and ±30°.

In the crack plane the value of $\psi$ is practically zero; therefore, in the line $\theta = 0°$ it was not measured. In the corresponding line $\theta = $ const, symmetric relative to the ray $\theta = 0°$, the values $\psi$ and $\varphi$ were averaged, and the calculations of $K_I$ were obtained from these averaged values.

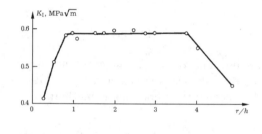

**Fig. 2.5**

Fig. 2.5, a typical graph, shows the value modifications of $K_I$ according to the coordinate $r/h$. It follows from the results obtained for each line $\theta = $ const that an interval can be selected where the value of $K_I$ is practically constant. The mid value of $K_I$ in the given interval, when $p = 2.23$ MPa, turned out to be equal to $0.58$ MPa$\sqrt{K}$. The deviations of mid values on different rays did not exceed 5% from the given value.

As an example using the results of the specimen-1 study, let us denote the near and the far boundaries of the asymptotic representation acting at the zone of stress near the crack tip by $r_1$ and $r_2$ respectively. It is clear that, for $0 \leq |\theta| \leq 90°$, the near boundary is situated at the distance of $r_1 \geq 0.7h$ from the crack. When $|\theta| \geq 90°$, the value $r_1$ increases to $1h$, i.e. the distance from the crack tip to the near boundary is a bit larger.

The far boundary of the indicated zone is $r_2 = 4h$ for $|\theta| \geq 90°$. When the angle $\theta$ changes from 90 to 30°, the distance $r_2$ is reduced from $4h$ to $2.8h$. The values $r_1$ and $r_2$ for $\theta = 0$ are obtained by extrapolation and are $r_1 = 0.7h$, $r_2 = 2.6h$ respectively.

The final form of validity zone of a two-dimensional stress state representation near the crack tip is shown in Fig. 2.6.

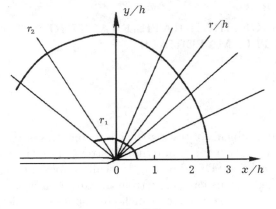

**Fig. 2.6**

The elongation of this zone in the $y$ direction to one and a half times its analogous value in the $x$ direction is very important and quite an unexpected result.

Thus, the studies carried out on two specimens of organic glass of different brands and thicknesses led us to the following results:

(1) the near boundary of the validity zone for a two-dimensional asymptotic stress-state representation is greater than or equal to $0.7h$ from the crack end. For $r < 0.7h$ an essential space-stress state is realised in the plate. The far boundary of this zone in the examined problem extends to a distance greater than or equal to $0.5\,l$ from the crack tip;

(2) the validity zone for a two-dimensional asymptotic stress representation has an unequal elongation in different directions from the tip and is more elongated in the direction perpendicular to the crack line.

The aforementioned results are very important in order to optimise the experimental determination of stress-intensity factors in zones with cracks in statics as well as in dynamics.

CHAPTER 3

# EFFECTS OF HIGH-RATE FRACTURE
# OF BRITTLE MATERIALS

One of the main problems of testing the characteristics of resistant materials
in dynamics is the dependence of dynamic strength on the way that the exterior action is applied. This difficulty typically appears under conditions of
high-rate loading. In this case, the strength can be interpreted as a critical
value of the stress-intensity factor which corresponds to microcracking near
the crack tip. The strength can also be interpreted as a dynamic local stress
leading to rupture continuum. Both are intensity limits of a local stress field
and the fracture occurs when these limits are reached. The dependence of
dynamic strength on the method of loading is manifested as critical values
during variations of action duration, of amplitude, and of rate of rise of the
exterior force. In the case of macrocrack motion initiation, such values will
be critical as regards the stress-intensity factor of growth of the macrocrack.
During fracture of 'intact' solids (i. e., not containing the given macroscopic
defects) the local cleavage stress is not determined by a material's characteristics but as a complex function of loading history.

## §1. Fracture of 'Intact' Materials

Experiments on fracture of 'intact' specimens demonstrate the dependence
of dynamic strength upon the velocity and loading duration even in materials characterised by almost ideal elastic–brittle behavior. This situation is
illustrated by the well-known diagram of the temporal dependence of the
strength, having been investigated first for metals by N. A. Zlatin et al. [8, 9].

Let us examine some of the principal results obtained from these experiments in terms of mechanics.

To create a controlled fracture during intensive pulse action with a duration of $\sim 10^{-6} \mu$ s, the cleavage phenomenon was applied. The methods, having been worked out at the AN RAN Physical-Technical Institute A. F. Ioffe,
made it possible to observe the history of a specimen's local loading in the
place of the cleavage-surface formation up to the moment of fracture.

The examined specimens were disk-shaped with a thickness of 25 mm. The
ramp loading was provided by a pneumatic gun. The pulse was elicited by the
impact of the bottom of an aluminum shell upon the central part of the disk
at an impact speed of up to 1050 m/s. Stresses, acting at the cleavage section, were determined according to interferometry measurements at the rear

surface of the specimen. These interferograms permitted a determination of the free-surface displacement speed in a reasonably continuous time interval following a number of sequential wave reflections in the cleavage plate. The stress time-dependence in the cleavage section was calculated according to the formula

$$\sigma(t) = \frac{1}{2}\rho c_1(V - V_S),$$

where $V$ is the free-surface velocity function; $V_S$ is the same function shifted in the argument on the interval, equal to the doubled time during which the longitudinal wave run through the cleavage-plate thickness; $\rho$ is the density; and $c_1$ is the velocity of longitudinal stress waves.

The method of estimating stress failure according to the 'dropping' velocity of a free surface under a temporal dependence which has been used in many papers (see, e.g., [11, 19]), is suitable only for perfectly trapezoidal pulses with a vertical forefront. This method cannot be mechanically transferred for example to triangular-shaped pulses. The aforementioned approach allows the formation of a full history of stress-state development at the cleavage section and is more efficacious as it can also be applied to arbitrary pulse forms.

Using this approach in experiments with aluminum and copper specimens (see [8, 9]), a dependence of the action time of tensile stress at the cleavage section on its amplitude value was found. These data were compared with the corresponding temporal strength dependencies obtained by quasistatic treatment. The results obtained from these two methods are quite different. Moreover the dynamic area of temporal strength dependence is practically independent of the static strength of the material. Experiments on aluminum and copper specimens revealed also that the location of the mentioned area remained invariable, even for a significant temperature variation. We note that in similar testing of the dynamic strength of a polymeric composite an opposite effect was obtained [4].

According to the data from dynamic tests, the cleavage fracture of materials at high rate by short-term threshold pulses occurs under stresses which exceed the static strength limit by many times. In this case, the time before a fracture is 'stabilised', meaning that the part of the dynamic diagram prior to the fracture is characterised by a weak time dependence on the amplitude of the threshold of the initial pulse. Similar tests with analogous results were carried out on a great number of specimens of different materials (see, e.g., [53, 19, 20]).

It was experimentally ascertained ([9]) that the location of the studied part on the temporal strength dependence diagram is not dependent on the static material strength. The results of these experiments show that it is not possible to accept the dynamic specimen rupture stress as a testing strength characteristic, because it is determined by a very strong dependence upon the parameters of the exterior action.

One significant observation revealed from this and other experiments is that during a fracture delay a macrofracture can occur as the local stress decreases. The physical nature of this effect has been discussed in many theoretical works (see, e. g., [35, 36, 61, 10]). However, the reason for this observation is not yet clear.

Altogether, we can conclude that the principal effects discovered during these cleavage experiments could not be explained by means of solid-medium mechanics, which are based on traditional notions about fracture.

## §2. Study of Dynamic Crack-Growth Resistance of a Brittle Medium

Similar problems as described for 'intact bodies' arise with attempts to characterise a material's resistance to dynamic crack growth. Experimental studies carried out at American research centers in the 1970s and 80s (see, e. g., [71, 74, 75, 78, 80, 82, 86, 115, 116, 118]) acquire a vital significance for a good understanding of fracture parameters under high-rate loading.

K. Ravi-Chandar and W. G. Knauss [83, 102–104] accomplished many experiments on impact loading of cracked specimens. There were a great number of effects accompanying crack initiation, fast growth, crack termination and branching. Thus, in these experiments they established rather amply the dependence of dynamic fracture durability on the loading history.

Experiments were conducted using specimens prepared with the vitreous polymer Homalite-100. The proportional pressure was created on the crack faces. This pressure increased linearly to a certain point and then remained constant. Loading was accomplished with the help of a special electromagnetic device which consisted of two copper strip-conductors with an electrically insulated central part placed inside the crack. A strong current pulse (of the order of $5 \times 10^4$ MA), was elicited by a discharge in an L. C. circuit which was fed to the plates. As a result of the electromagnetic action, pressures appeared which separated the plates and contracting stresses distributed along the crack surface.

Their temporal dependencies were determined by the parameters of the L. C. circuit. In particular, trapezoidal pulses were used with the fixed stress-increase time $t_0$ equal to $25\,\mu$s, lasting $\sim 160\,\mu$s. Maximal pressure on the crack surface, equal to 62 MPa, was obtained at the loading rate $2.5 \times 10^6$ MPa/s.

The stress field around the crack tip was investigated by the method of caustics with the use of high-speed photography. The frequency of the camera reproductions used was equal to $2 \times 10^5$ exposures/s and the exposure time was $20\,\mu$s. By using high-speed photography, temporal dependencies were obtained for the stress-intensity factor and for geometrical crack characteristics. The initiation stress-intensity factor was obtained by interpolation of the data at the moment of initiation of crack growth as determined by the history of crack development.

The dimensions of the specimens were chosen in such a way that reflected waves would not have time (of order of $150\,\mu$s) to interact with the crack tip, so this imitated an unbound medium. The thickness of the specimens was 4.8 mm. The reliability of the method of caustics and the system's ability to produce similar loading pulses were investigated by experimenters ([102]) beforehand.

The main advantage of the described experimental device is its simplicity of synchronisation. The loading program is easily coordinated with the launching moment of the high-speed camera, and it becomes possible to register on the film the crack behavior at launch.

The main result of these experiments is that the critical value of the stress-intensity factor $K_{Iq}$ increases with increasing load rate and can significantly exceed the corresponding quasistatic value (Fig. 3.1). Moreover, it turned out that the influence of loading rate on the launching value of the aforementioned coefficient can be neglected, if the time before fracture is $t_* \geq 50\,\mu$s; it corresponds to the loading velocity $-10^4$ MPa/s. With a loading rate increase, the corresponding time before specimen fracture decreases, and the critical value of the stress-intensity factor notably increases.

The researchers note that if the time is about $\geq 50\,\mu$s the fracture is determined by a quasistatic criterion of a critical stress-intensity factor, which is not dependent on time or velocity.

For lesser times or for high-rate loading new approaches ought to be used for estimation of the possibility of a fracture.

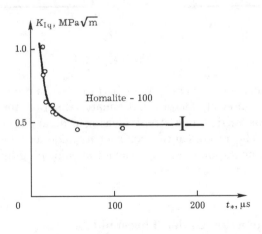

Fig. 3.1

As a working hypothesis K. Ravi-Chandar and W. G. Knauss have supposed that under the high rates of loading the dependence of the behavior of the material itself becomes significantly strain-rate sensitive. In this case the material had to be characterised as non-linear and viscoelastic. Accordingly, it is stated that under smaller rates of loading an equilibrium behavior for a material is common, and under higher rates it is the 'viscous' reaction that dominates. Experimentally obtained data could be approximated by a power dependence

$$K_{\mathrm{Iq}} = K_{\mathrm{Ic}} + \frac{C}{t_*^2}.$$

If we suppose the necessity of attainment of the critical opening displacement at the crack tip, this must conform to the creep law in the form of $D(t) = D_0 t^r$, $r = 2$. Let us note that $r = 2$ is usually accepted for polymer creep under small deformations.

Theoretical graphs of $K_{\mathrm{I}}(t)$ change for semi-infinite (a) and finite (b) cracks as shown in Fig. 3.2. The moment of fracture in the described experiments turns out to be on the escalating branch of the graph b. For example, in the theoretical analysis of this case we can use solutions for a semi-infinite crack.

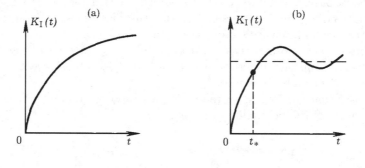

Fig. 3.2

These experiments convincingly demonstrate that in dynamics the critical value of the stress-intensity factor is not a material parameter, and, therefore, attempts to measure dynamic strength using ordinary static methods are quite erroneous. The authors of the experiments under consideration note the impossibility of simulation of crack behavior at launching by the traditional methods of continuum mechanics.

## §3. Crack Behavior Under Threshold Loading

Experiments published in a number of works by the Stanford Research Center International (SRC International) are of great importance for understanding high-rate fracture when the possibility of unstable macrocrack growth under threshold exterior loading was considered. J. F. Kalthoff and D. A. Shockey

[78] were the first to measure threshold amplitudes of short pulses for disk cracks. Then similar experiments were continued by other researchers ([73, 112]).

Let us consider the most salient features and conclusions from the afore-mentioned works.

Suppose a specimen having a crack of a length $L$ is subjected to an impact action and the result is a short-term wave pulse of stress interacting with the crack. Let this pulse of a duration $T$ and an amplitude $P$ have a rectangular form. This should allow a determination of the dependence of the critical crack dimension on the applied pulse amplitude.

An alternative for the same problem can be presented as a determination of the minimum destroying amplitude (threshold) of the applied effort with the given pulse duration and fixed geometrical parameters of the specimen. In statics, the examined problem can be solved by the classical Griffith–Irwin method. There is a series of experimental methods corresponding to various situations allowing the study of quasistatic threshold characteristics of fracture (see, e.g., [59, 44]).

If we apply a short-term pulse of dynamic action to the specimen, the situation will be more complicated. From the data of dynamic tests it is known that high rates of loading change the fracture toughness: $K_{\mathrm{Id}}$ essentially differs from $K_{\mathrm{Ic}}$. Complicated diffraction interactions, appearing near the crack, lead to oscillations of the stress-intensity factor. Altogether, the analysis becomes extremely complicated by mathematical difficulties as well as by an absence of adequate fracture criteria.

J. F. Kalthoff and D. A. Shockey [78] have carried out an experimental investigation of this problem with polycarbonate specimens. A plate, fixed on a cylinder shell, was accelerated with the help of a pneumatic gun and struck against a plate target. The time of the pulse growth in the target was 10–100 nano s. Its duration was equal to several microseconds. The pulse amplitude and its duration are simple functions of the striking velocity and of plate thickness. They are easy to control, to compute, or to measure directly.

This experimental approach permits, firstly, the creation of a great number of cracks in the target and, secondly, a test of their stability to further impacts. It was shown that the criterion of unstable crack germination, so traditional in quasistatics, can not be used in dynamics. This is especially so if the time of loading is comparable to the time of the wave run at a distance equal to the crack length. In this case the stability threshold, calculated according to the classical criterion of the stress-intensity factor, turns out to be significantly smaller than the value obtained by experimentation. The results of these studies are shown schematically in Fig. 3.3.

H. Homma et al. [73] have experimentally investigated the same problem on specimens of different configurations, made from different materials. Tests were accomplished on specimens made from two kinds of steel (steel 4340 and steel 1018) and aluminum (aluminum 6061–T651). The fatigue macrocrack

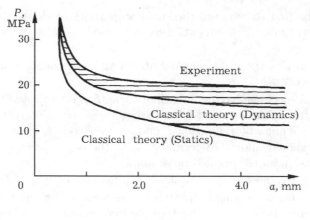

Fig. 3.3

was cultivated from a form of wide strips with a boundary cut in them. Short pulses with a given duration and amplitude were created by cylinder shells and a special compressor device. The minimal amplitude, corresponding to fracture of all three specimens, was determined. Also fractures caused by large-amplitude stresses were investigated. The threshold amplitude was found by means of incrementally increasing the impact velocity to the point where initial crack propagation was launched. The fracture, fixed for the first time, advanced at a rate of $10$–$50\,\mu$m. The amplitude was increased by small increments. Its critical value was obtained as a median between the nearest destroying and non-destroying amplitude values. The influence of pulse duration upon the threshold amplitudes was also analysed. For this purpose cylinder shells of different lengths were used. As a result of these tests, the dependencies of critical amplitudes upon the initial crack length were constructed, and the relation between the length of the crack 'skip' and the overload, i. e. exceeding of the threshold amplitude, was investigated.

Tests on specimens from many materials have confirmed the decrease of critical values of amplitudes $P$ according to the increase of the crack length $L$ and the threshold growth under the shortening of pulse duration. These results were compared with the static dependence, given by the formula

$$P = \frac{K_{\text{Ic}}}{\sqrt{L}}\, F\left(\frac{L}{W}\right),$$

where $W$ is the specimen relative width. For short cracks (long pulses) the form of curves under dynamic and static loading are alike, but for long cracks (short pulses) the values, measured during the tests, tend to a finite horizontal, and are located essentially higher than the static curve (see Fig. 3.3), tending toward zero.

The form of the experimental threshold curve is qualitatively explained and forecasted with the help of the initial boundary value and an analysis of the corresponding dependence of the stress-intensity factor upon time. For short cracks, the fracture happens as in statics. In this case the time of loading includes a great number of the mentioned coefficient-value oscillations with its exit to a static value. The destroying stress must decrease when $L$ increases. The threshold amplitude for large cracks depends on the stress intensity factor variation with time. If the crack length exceeds some critical value, the dependence of threshold amplitudes upon $L$ disappears. In this case there are no oscillations of the stress-intensity factor values, as the wave has no time to overcome the distance, corresponding to the crack length. The threshold amplitude on the graph $P - L$ turns out to be constant (see Fig. 3.3).

Nevertheless, the attempts undertaken to compute threshold fracture characteristics have turned out to be unsuccessful. According to classical approaches ([114, 62]), a crack, loaded by a critical pulse, advances under the condition that the dynamic stress-intensity factor reaches its maximum. The corresponding limit relation, whence threshold amplitudes can be determined, is written as

$$\max_t K_{\mathrm{I}}(t) = K_{\mathrm{Ic}}. \tag{3.1}$$

Analysis has shown that for short pulses (long cracks) experimental critical amplitude values significantly exceed values calculated theoretically according to (3.1).

The analysis of the experimental data and the computation of results for the indicated specimens has shown ([73, 112]) that the fracture around the crack tip can also happen on the falling curve line, corresponding to the stress-intensity factor change (effect of fracture delay), that is also unexplainable if we use traditional fracture notions.

# CHAPTER 4
## SOME NON-CLASSICAL APPROACHES.
## NEW CRITERION OF BRITTLE FRACTURE

Many modeling methods of dynamic fracture effects are associated with non-elastic rheology and macrocrack development. In many cases this association is an inevitable physical necessity. However, for practical aims it is very important to provide a direct mechanical approach, permitting the reduction of the dynamic fracture analysis to a simple 'industrial' procedure. That is why the rejection from the engineering mode of energy and power balance traditional schemes of fracture mechanics would be unjustified. Even in the framework of linear elasticity and brittle fracture these schemes are not complete. Their development, as will be shown below, can give sufficiently simple explanations of many peculiarities of high rate-fracture [23, 25, 26, 30, 31, 45, 47, 99].

Let us examine some nonclassical modeling methods of brittle-material fracture, especially efficient in situations where the classical approaches and the Griffith–Irwin criterion do not ensure success.

## §1. On the Neuber–Novozhilov Criterion

Let us turn to static problems. We will consider an elastic plate with an angular notch. According to Griffith's classical scheme we compose an energetical balance equation.

We have $\Delta\Pi \sim \varepsilon$. An estimation for $\Delta(A - U)$ is also known ([17])

$$\Delta(A - U) \sim \varepsilon^{2\pi/\alpha}.$$

It is evident that we would not able to find critical efforts from the equality

$$\Delta(A - U) = \Delta\Pi.$$

Let us study two elastic plates, weakened by a rectilinear crack and a lune of small apex angle $\alpha$. In the first case near the crack tip we have an asymptotic expression

$$\sigma_{ij} = \frac{C^{\mathrm{I}}_{(1)}}{r^{\lambda^{\mathrm{I}}_1}} \, f^{\mathrm{I}}_{ij(1)}(\theta) + \frac{C^{\mathrm{I}}_{(2)}}{r^{\lambda^{\mathrm{I}}_2}} \, f^{\mathrm{I}}_{ij(2)}(\theta) + \dots,$$

where $\lambda^{\mathrm{I}}_1 = \lambda^{\mathrm{I}}_2$; $C_1, C_2$ are coefficients characterising the stress-state intensity; and $r$, $\theta$ are coordinates near the crack tip.

In the second case the roots are separated

$$\sigma_{ij} = \frac{C_{II}^{(1)}}{r^{\lambda_1^{II}}} f_{ij(1)}^{II}(\theta) + \frac{C_{II}^{(2)}}{r^{\lambda_2^{II}}} f_{ij(2)}^{II}(\theta) + \dots,$$

where $\lambda_1^{II} < \lambda_2^{II}$.

According to Irwin's criterion, in the second case only the dominating term should be taken into account; this leads to a significant difference between the situations I and II that is difficult to interpret. H. Neuber [33] and V. V. Novozhilov [38, 39], at different times and on the basis of different approaches, have suggested the following fracture criterion

$$\frac{1}{d} \int_0^d \sigma \, dr \leq \sigma_c. \tag{4.1}$$

Here $\sigma$ is the main tension stress near the crack tip ($r = 0$); $\sigma_c$ is a stress limit of 'intact' material.

The main peculiarity of (4.1) is an introduction of some structural dimension $d$ in an explicit form. We note that a structural characteristic of the length dimension is already implicitly present in classical fracture mechanics, appearing in the form of dimensional combinations of the classical strength criterion parameters

$$d \sim \frac{\Gamma E}{\sigma_c^2}, \qquad d \sim \frac{K_{Ic}^2}{\sigma_c^2},$$

$E$ being Young's modulus, $\Gamma$ the specific surface energy and $\sigma_c$ the critical stress.

Scholars express various suppositions according to the physical nature of the parameter $d$ (interatomic spacing for a medium with a regular atomic structure, grain size for a polycrystalline medium, a scale correspondence parameter of strength characteristics, etc.). We propose considering the parameter $d$ as a linear dimension, characterising the fracture elementary cell on the given scale level (see, e.g., [5, 6]). Without giving any details we choose $d$ from the condition

$$d = \frac{2K_{Ic}^2}{\pi \sigma_c^2}, \tag{4.2}$$

providing coincidence of (4.1) with the Griffith–Irwin criterion in 'simple' cases.

Criterion (4.1), in combination with (4.2), permits efficacious fracture forecasting in many nonstandard situations, including the aforementioned cases of plate fractures with angular and lune notches ([22]).

## §2. The Shockey–Kalthoff Minimal-Time Criterion

The testing of dynamic load pulse threshold characteristics, considered in the preceding chapter, has permitted the authors of those experiments to draw a conclusion on the necessity of revision of traditional quasistatic fracture mechanics notions.

J. F. Kalthoff and D. A. Shockey [78] have suggested a new fracture criterion, which they call the minimum-time criterion. The main novelty of the new approach is an introduction of a structural parameter $t_{inc}$, having a time dimension and accounting for incubation processes preceding macrofracture. Incubation time $t_{inc}$ is declared to be a constant, linked to material properties. According to this concept, fracture occurs on the condition that the stress-intensity factor current value $K_I(t)$ exceeds the fracture dynamic viscosity $K_{Id}$ during some minimal time needing for macrocrack development.

In the case of threshold pulses an unstable crack development happens if $K_I(t) \geq K_{Id}$ during a time equal to $t_{inc}$. Here the fracture dynamic viscosity $K_{Id}$ is determined according to a quasistatic formula, e.g. $K_{Id} = 2P\sqrt{a/\pi}$ for a disk crack, where $a$ is the crack radius, and $P$ is the amplitude of the corresponding dynamic stress pulse, under which the fracture occurs. Incubation-time values $t_{inc}$ were determined for different materials [73, 112], in particular, for steels 4340 ($7\,\mu s$), 1018 ($11\,\mu s$) and aluminum alloy 6061–T651 ($9\,\mu s$).

Notwithstanding some eclecticism and absence of a neat analytical setting, the minimal time criterion is a notable step forward, as it introduces the following principal directives in fracture analysis.

Firstly, the existence of a certain structural parameter, having the time dimension and controlling the fracture process is considered. Let us remark that in quasistatics the fracture is associated with a certain parameter having a length dimension. Thus, during the passage from static loading to a dynamic one a new structural characteristic appears.

Secondly, it is affirmed that the fracture is not stipulated by momentary states of the local stress field near the crack tip, but is an integral process, existing in time, and distributed at a structural–temporal interval.

The minimal-time criterion admits the possibility of the existence of such a situation, which may be unusual from the traditional point of view, when fracture occurs at the temporal segment of a local stress-field intensity decrease. In fact, from the point of view of classical mechanics, if a fracture does not happen when the stress-intensity factor attains its maximal values, then it can not happen for smaller values of the stress-intensity factor. The minimal-time criterion, on the contrary, accepts well such a possibility. As was already noted, experiments on threshold pulses proved the existence of the given effect, that could be called an 'effect of fracture delay', by analogy with a similar effect for a cleavage.

## §3. The Nikiphorovski–Shemyakin Pulse Criterion

An important principle of fracture modeling in dynamics was advanced by V. S. Nikiphorovski and E. I. Shemyakin [35]. It is direct accounting of the local rupture stress history. This criterion corresponds to the attainment by the complete integral of the local tension stress taken over time, i. e. the complete local force pulse compared with a critical value, determined from experiments

$$\int_0^{t_*} \sigma \, dt \leq J_c. \tag{4.3}$$

This criterion was specially suggested for fracture analysis, provoked by short-term exterior pulses. It permits a qualitative explanation of the strength augmentation (critical rupture stress augmentation) of materials at elevated loading velocities.

It is interesting that (4.3) itself could be interpreted from the phenomeno-logical theory of defect accumulation ([12]). Let us suppose the following phenomenological law of defect development to be valid

$$\frac{d\Psi}{dt} = f(\sigma, \Psi) = \begin{cases} A \left( \dfrac{\sigma}{1 - \Psi} \right)^n, & \sigma \geq 0, \\ 0, & \sigma < 0. \end{cases} \tag{4.4}$$

Integrating (4.4), provided that $\Psi \leq 1$, we have

$$A \int_0^{t_*} \sigma^n \, dt \leq 1,$$

which, for $n = 1$, gives (4.3).

Despite possible interpretations, (4.3) is a postulate, determining macro-rupture of an 'intact' solid medium. Its main deficiency is the impossibility of a transition to quasistatics. Thus, according to (4.3) any stress, even insignif-icant, of the form of $\sigma = \sigma_0 H(t)$, where $H$ is the Heaviside function, leads to fracture. It makes the location of a dynamic branch on the strength/temporal dependence curve indeterminable, permitting the interpretation of some tem-poral effects of dynamic fracture only in a qualitative way.

The introduction of pulse criterion (4.3) was an important step forward, as it provided a possibility of direct consideration of loading history, through the fracture mechanism. Such an approach, in the framework of the elastic–brittle model, gave an explanation of a few principal effects of fast dynamic rupture of solids ([35]).

## §4. The Structural–Temporal Criterion of Brittle-Medium Fracture

Analysis of experimental results shows that the main contradictions with data, obtained from traditional models, become apparent in the case when the fracture happens in rather short time intervals after the beginning of exterior pulse application, which correspond to high loading rates. The fracture itself is accompanied by high local deformation velocities, both during cleavage and the initiation and fast growth of cracks. The indicated contradictions may appear because the fracture modeling, used for analysis, remains essentially based on quasistatic principles.

Thus, in the modeling of a fracture mechanism it is not taken into consideration that during the high-rate rupture, together with elastic resistance of the material, it is also necessary to overcome the medium inertia. In the case of application of an energetic balance equation to the dynamic fracture the term, corresponding to the kinetic energy, is traditionally rejected, as being small compared with other terms ([42, 43, 69–71]).

Evidently, this approach is incorrect for a high-rate deformation. The force criterion, corresponding to this approach, is formulated as a requirement of attainment of the stress-intensity factor, of a certain critical value, determined by material testing.

The physical imperfection of this criterion is that, according to it, the material must be destroyed at a rather high instantaneous local force, acting at the crack tip. In fact, a high mobility of small but extremely fast medium particles determines the dependence not only on instantaneous components of the force field but also on the time of their action.

One more direction of classical modeling development lies in the selection of fracture process structural characteristics. Nowadays, the necessity of structural parameter accounting is not brought in to question, as, practically, all the experimental results testify in its favor. The only question is what characteristics of microstructure physical processes have to be introduced in the macromechanics of fracture. However, this selection is quite complicated and delicate, as it is very important not to go beyond the bounds of reasonable detail. In our opinion, the most natural would be a search of such parameters, as the Neuber–Novozhilov dimension in quasistatics, being universal and not limited by unequivocal physical interpretation, which could be different according to material peculiarities, scale level and problem class.

Referring to the classical force approach of fracture mechanics, we notice that it has the sense of attaining a critical value, by an instantaneous local force acting in the supposed place of rupture. However, in dynamics one should consider inertia, since the medium particles, adjacent to the rupture place, can move extremely fast. At the same time, in analogy with the structural dimension, known in statics, it is natural to consider the structural time in dynamics. So, in the simplest case, for structural dimension $d$ and maximal wave velocity $c$, the relation $d/c$ is the characteristic time of interaction energy transmission between adjacent elements of the fracture structure.

Thus, let us suppose we have a characteristic time interval $\tau$, corresponding to an incubation (latent) period of macrofracture development. Generally, the parameters $d$ and $\tau$ should be evaluated as independent, as $\tau$ could be determined by complicated processes, occurring in the material structure.

We introduce an elementary spatial–temporal fracture cell $[0, d] \times [t - \tau, t]$ (Fig. 4.1). In other words, we suppose that a certain structure, characterising peculiarities of the fracture process on the prescribed scale level, is set on the spatial–temporal scale.

Assume that the fracture happens, if we have an equality in the following condition

$$\frac{1}{\tau} \int_{t-\tau}^{t} dt' \frac{1}{d} \int_{0}^{d} \sigma(t', r) \, dr \leq \sigma_{\mathrm{c}}, \tag{4.5}$$

where $\tau$ is the fracture structural time; $\sigma_{\mathrm{c}}$ is the static strength of the material; $\sigma(t, r)$ is the tension stress near the crack tip $(r = 0)$; and $d$ is a length scale parameter in correspondence with strength characteristics, determined according to data of static tests on specimens with macrocracks. In this case (4.5) coincides with the Neuber–Novozhilov criterion, uniquely connecting $d$ with static characteristics of fracture viscosity and material strength (4.1).

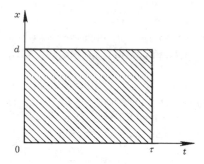

**Fig. 4.1**

In accordance with (4.5), the maximal tension stress near the crack tip, averaged over the spatial–temporal interval $[0, d] \times [t - \tau, t]$, must attain the strength of the material. By introducing new notations, we rewrite (4.5) in the form

$$J(t) \leq J_c, \quad J(t) = \int_{t-\tau}^{t} dt' \int_{0}^{d} \sigma(t', r) \, dr, \quad J_c = \sigma_c \tau d.$$

Thus, the introduced criterion relation has the physical meaning of a critical structural pulse.

The main problem of the dynamics is to determine the moment of fracture. In the considered case it is natural to consider the fracture time $t_*$ as a moment when the given pulse attains its critical value: $J(t_*) = J_c$.

While determining the loading parameters, corresponding to the minimal fracture charges, an important role is played by the threshold condition

$$J(t_*) = \max_t J(t) = J_c,$$

that can be considered as an optimality condition and a condition of unstable crack growth.

According to this approach $\sigma_c$, $K_{Ic}$ and $\tau$ form a system of defining parameters (in the simplest case – of constants), reflecting strength properties of the material.

Further, it will be shown that the spatial–temporal criterion gives an opportunity to avoid an a priori introduction (measurement) of 'material' functions of dynamic strength the material and to consider, e.g., the dynamic fracture viscosity, dependent on rate, duration and other exterior loading characteristics, as an estimated problem performance.

## §5. On the Discrete Nature of Dynamic Fracture of Solids

We will show that the principle of tension stress field critical intensity, traditional in continual mechanics, does not cohere with the law of variation of momentum.

Let the material rupture in a one-dimensional cleavage problem ([35]) be caused by a triangular-profile stress pulse with duration $T$. Let us determine the threshold, i.e. the smallest force fracture pulse $U = U_c(T)$ for the given $T$. Using the classical critical stress criterion $\sigma \leq \sigma_c$, we obtain

$$U_c = \frac{1}{2}\sigma_c T.$$

The corresponding threshold is presented in Fig. 4.2 (hatched line). It is obvious that the fracture domain on the diagram corresponds to points lying above the threshold curve, adjoining the origin of coordinates. Hence, even infinitesimal pulses are capable, according to the accepted criterion, of causing fracture.

We study the problem of a semi-infinite crack for an unbounded plane. Let a uniformly distributed shock stress of temporal rectangular profile act on the crack surfaces $x \leq 0$, $y = \pm 0$

$$\sigma_y = P\Big[H(t) - H(t-T)\Big], \quad \sigma_{xy} = 0,$$

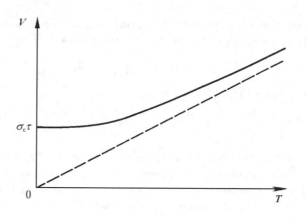

**Fig. 4.2**

where $P$ and $T$ are amplitude and pulse duration. Then on the crack plane we get

$$\sigma_y = \frac{K_I(t)}{\sqrt{2\pi r}} + O(1),$$

where $K_I(t)$ is expressed by (1.1). Using the classical criterion of critical stress intensity factor $K_I \leq K_{Ic}$ we have

$$U_c = \frac{K_{Ic}\sqrt{T}}{\Phi(c_1, c_2)}.$$

Therefore, for $T \to 0$ the fracture threshold pulses become infinitesimal.

The conclusion drawn contradicts common sense, and the well-known experiments on high-rate fracture, as was shown beforehand, demonstrate various effects that could not be placed in the framework of traditional approaches.

According to classical criteria, under dynamic rupture of a material the energy and the pulse, consumed for formation of new surfaces and domains of fracture, are spent incessantly. We will show that an elementary account of dynamic rupture process physical discreteness leads to a structural–temporal criterion of fracture.

It has already been mentioned that the main parameter of crack mechanics is a linear dimension $d$, characterising an elementary fracture cell. Such a cell has no unambiguous physical interpretation for all practical cases and is, in fact, a universal fracture characteristic. It can be interpreted in different ways according to the class of studied problems.

We introduce a pulse elementary portion ('quantum'), required to destroy one structure cell: $U_1 = \sigma_c \tau$; here $\tau$ is the incubation time, determined by the material properties and the class of problems.

We suppose that under cleavage a threshold pulse of given shape, e. g. triangular or rectangular, with duration $T$ is created in the medium, with the result that there is a fracture of a certain number of structural elements. The fracture of $m$ structural cells requires the pulses

$$U_m = \sigma_c \tau m, \qquad m = 1, 2, 3, \ldots$$

Let us introduce the distribution

$$P_m = C \exp\left(-\frac{U_m}{\alpha T}\right), \tag{4.6}$$

where $P_m$ is the probability of fracture of $m$ structural cells; $\alpha$ is a parameter, depending on the shape of the temporal stress profile and obtained from the exit condition to quasistatic characteristics for sustained loading; and $C$ is a normalising multiplier, determined by the relation

$$\sum_m P_m = 1. \tag{4.7}$$

The average threshold pulse can be found by the formula

$$U = \sum_m P_m U_m. \tag{4.8}$$

From (4.6)–(4.8) for a triangular pulse we have

$$U = \frac{\sigma_c \tau}{1 - \exp(-2\tau/T)}.$$

The corresponding threshold is shown in Fig. 4.2 by a continuous line. Evidently, threshold pulse finite values correspond now to short times. Under sustained loading ($\tau \ll T$) the threshold characteristics can be computed according to the classical critical stress criterion

$$U = \frac{1}{2}\sigma_c T.$$

Let us determine a macrorupture as a fracture of at least one structural element ([38]). Then, the corresponding criterion can be written in the following form

$$\int_{t-\tau}^{t} \sigma(t')\,dt' \le U_1 \equiv \sigma_c \tau. \tag{4.9}$$

Here $\tau$ is the minimal time, necessary for a local threshold pulse with a duration that will cause a macrofracture.

For media with sharp concentrators, such as cracks, the mean values of rupture stress on the structural interval are examined. Instead of (4.9) we have

$$\int_{t-\tau}^{t} dt' \int_{0}^{d} \sigma(t', r)\,dr \le \sigma_c \tau d, \tag{4.10}$$

coinciding with (4.5). In a particular case of quasistatic loading (4.10) coincides with the force criterion of Neuber–Novozhilov. The received limit condition practically coincides with the already introduced structural–temporal criterion. In this case (4.9), in the absence of sharp concentrators such as cracks, can be considered as a particular case of (4.5).

Now, according to (4.9) and (4.10), the dynamic strength of a brittle medium can be evaluated as a calculated characteristic. Moreover, it is natural to expect that both the critical rupture stress of 'intact' continuum and the fracture viscosity of cracked domains will show a dependence on parameters of the exterior action, including the rate of loading. We have established that such a behavior is the principal singularity of dynamic fracture, stipulated by the 'quantum' nature of this process.

Essentially, the analysed problem of dynamic strength under a high-rate loading is an analog to the problem of low-temperature heat capacity of solids for classical molecular physics ([13]). This problem was solved from the position of quantum mechanics. The postulates on the discrete substance structure (a solid is a combination of elementary oscillators), on the discrete nature of energy liberation and absorption (energy is liberated and absorbed by elementary portions – quanta) and correspondence principle (in the limits of low load rate cases the quantum theory should not conflict with the classical one) were taken as a basis. The presented considerations permit us to overcome the classical theory difficulties and to explain the dependence of specific heat capacity of solids on temperature. In this case it turned out that for the lowest temperatures (close to absolute zero) the solid energy is finite and is determined by the elementary quantum energy, and the corresponding temperature dependence of heat capacity can be calculated fairly easily.

The dependence of solid interior energy $\langle E \rangle$ on temperature $\Theta$, being calculated according to quantum (continuous line) and classical (dashed line) theories is shown in Fig. 4.3.

Analogy between essential notions of quantum mechanics and basic principles of V. V. Novozhilov's theory is evident:

(1) all solids consist of spatial–structural elements of finite size;
(2) an elementary act of fracture is a fracture of one structural element;
(3) criterion parameters, including a structural element dimension, should be chosen in such a manner that in the limit of low load rates the obtained classical fracture theory results should be preserved.

This analogy becomes even more apparent if we compare the dependence $\langle E \rangle$–$\Theta$, presented in Fig. 4.3, with the dependence of a threshold (minimal destroying) force pulse on its duration under cleavage (see Fig. 4.2). The similar dependencies for cleavage strength and dynamic crack stability will be obtained further on the basis of the structural–temporal criterion. Evidently, neither the strength for high velocities under low temperatures, nor the heat capacity under low temperatures, can be modeled on the basis of continuum notions.

As in the case of heat capacity, the concept of dynamic fracture 'quantum' nature and corresponding criteria (4.9) and (4.10) permit us to avoid an a priori introduction of 'material' functions, determining the material dynamic strength, and to examine them as additional characteristics of the problem.

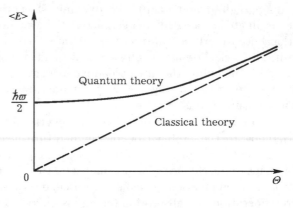

**Fig. 4.3**

Globally, the aforementioned analogy with the behavior of solid heat capacity at low temperatures has something in common with the well-known principle of temperature–temporal (or temperature-rapid) correspondence of deformable solid mechanics: both for low temperatures and high rates the discrete ('quantum') character of a process gets a decisive significance.

The idea of fracture process discreteness has been discussed on repeated occasions in scientific works. Thus, ideas of substitution of a solid medium by discrete geometrical structures permit important conclusions to be drawn on the brittle equilibrium nature for limit values of static and dynamic efforts (see [38, 39, 119, 54, 22, 32, 92]).

The general particularity of these ideas is the discreteness on the geometrical level, i.e. a fracture analysis of some discrete construction (e.g. chain or lattice) on the basis of quasistatic determining characteristics. The rupture criteria, used for such an approach, remain the same, i.e. they are borrowed from the static continuous fracture mechanics. Attainment by the local tension stress of its critical value can be taken, e.g., as a natural criterion of lattice rupture.

It has been established above that a direct transference of this principle to dynamic processes was incorrect. Analysis of fracture dynamics requires the following second step: a quantisation at physical level, i.e., an introduction of energy (pulse) consumption discreteness, necessary to support the fracture process. Such a juncture is typical for physics, and it can be repeatedly observed under the transition of process load rates from moderate to extremal ones.

## §6. On the Relaxation Nature of the Incubation Time

We will show that the considered structural–temporal criterion is closely connected to the relaxation processes, accompanying rupture development in continuum.

We suppose that a given point of the material is characterised by an intensity of a stress field $\Sigma(t)$, with the result of deformation and development of microdamages. Let the following deformational-based fracture criterion be valid under these conditions

$$K\chi(t) \leq \Sigma_{\mathrm{c}}, \qquad (4.11)$$

where $K$ and $\Sigma_{\mathrm{c}}$ are material constants, and $\chi(t)$ is a relative volume modification, caused by deformation and microdamage in the given point.

If the material is linear-elastic, then $\Sigma(t) = K\chi(t)$, and from (4.11) we obtain an analogous critical stress criterion

$$\Sigma(t) \leq \Sigma_{\mathrm{c}}.$$

Now, let the material be subjected to the rheological law

$$\Sigma(t) = K\chi(t) + \mu\frac{\mathrm{d}\chi}{\mathrm{d}t}, \qquad (4.12)$$

where $\mu$ is a viscosity factor. Let the temporal loading profile be given. Solving (4.12) with respect to $\chi(t)$, we get

$$\chi(t) = \frac{1}{\mu} \int\limits_{-\infty}^{t} \exp\left[-\frac{K}{\mu}(t-s)\right] \Sigma(s)\,\mathrm{d}s. \qquad (4.13)$$

The kernel of the integrand of (4.13) is the function $\exp[-(K/\mu)t]$. We replace it by a step function $\theta(t)$ (Fig. 4.4) in such a manner that

$$\int\limits_{0}^{\infty} Q(s)\,\mathrm{d}s = \int\limits_{0}^{\infty} \exp\left(-\frac{K}{\mu}s\right)\,\mathrm{d}s = \frac{\mu}{K}.$$

Then (4.13) is converted into the relation

$$K\chi(t) = \frac{1}{\mu/K} \int\limits_{t-\mu/K}^{t} \Sigma(s)\,\mathrm{d}s,$$

whence, considering (4.11) and using the notation $\tau = \mu/K$, it follows that

$$\int_{t-\tau}^{t} \Sigma(s)\,\mathrm{d}s \leq \Sigma_c \tau. \tag{4.14}$$

Condition (4.14) coincides completely, in form, with the structural–temporal criterion (4.9) for 'intact' materials.

Small values of a 'viscous' term in (4.12) correspond to small viscosity and long-time deformation

$$\frac{\mu}{K}\frac{d\chi}{dt} \ll 1.$$

In this case the critical stress criterion is valid. The 'viscous' term should be taken into account for high-rate dynamic loading, and (4.14) must be used for fracture estimation.

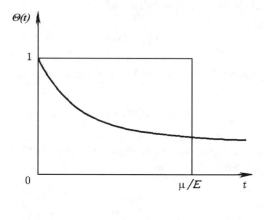

Fig. 4.4

The obtained temporal characteristic $\tau = \mu/K$ has the physical meaning of relaxation time. However, it should be kept in mind that the real relaxation is caused not only by (and even not as much as) viscous deformation, but is a result of microfracture, preceding a macrorupture of the material. L. Seaman et al. [109, 110] have shown that dynamic macrofracture and change of microvolumes of brittle materials, accompanying it, could be described by an equation of the form of (4.12), and the relaxation times of brittle steels and alloys, corresponding to this process, turn out to be larger by several orders to analogous characteristics of simple viscous deformation of these materials.

Thus, the macrofracture analysis, being made on the basis of the structural–temporal criterion, underlines the cardinal importance of study of physics of relational processes, accompanying the irreversible deformation and formation of microdefects ([123]), as well as the determination of corresponding relaxation time. In particular, the presence of the relaxation time's spectra can be a premise to categorise structural fracture levels on a temporal scale.

## §7. On Choice of Parameters of the Structural–Temporal Criterion

The choice of appropriate fracture criterion parameters has a basic influence on the possibilities to obtain quantitative characteristics for specific problems. On solving problems on a macrolevel, this choice is determined by the investigator's arbitrariness. Only experimental results can confirm its correctness.

The structural–temporal fracture criterion, given supra, is based on a system of three determining parameters: $(\sigma_c, K_{\mathrm{Ic}}, \tau)$, two of which – static strength and static viscosity of fracture (characterising the structural dimension $d$) – are well known. The third parameter, representing the structural (incubation) fracture time, could be interpreted in different ways on the microphysical level, but, in the end, it must be postulated by and specifically chosen.

Let us consider two main possibilities of such a choice

(1) the structural time $\tau$ is determined with the help of the fracture structural dimension

$$\tau = \frac{d}{c} = \frac{d\sqrt{\rho}}{k}. \tag{4.15}$$

Here $c$ is the maximum wave velocity; $\rho$ is the density; and $k$ is a constant, depending on the deformation properties of the material. According to this determination, the fracture structural time has the physical meaning of average transmission time of energy interaction between adjacent elements of the fracture structure.

Further, it will be shown that the structural–temporal criterion with parameter $\tau$, obtained according to (4.15), permits efficient calculations of dynamic strength characteristics of 'intact' materials. The results of dynamic strength calculations on the basis of (4.15) are well justified by the data of cleavage experiments [8, 9, 19, 20];

(2) the incubation time $\tau$ does not directly depend on the fracture structural dimension and has to be obtained experimentally, e. g. under fracture initiation near a tip of a macrocrack. The conceiving, growth and confluence of numerous microdefects in a certain (sufficiently large) area in the neighborhood of the crack tip, preceding a macrorupture of a material, determine the characteristic scale level of macrorupture. The incubation time could be considered as some integral temporal characteristic of these processes. Further it will be es-

tablished that under the formation of macrocracks the fracture structural time $\tau$ can be interpreted as the incubation time $\tau = t_{\text{inc}}$ from the known minimal-time criterion, as suggested in [73, 78, 112].

Evidently, the two considered ways of choosing $\tau$ do not exhaust all the possibilities of this characteristic interpretation. Without contesting the interpretation importance of introduced structural values $d$ and $\tau$ (which can change according to situation) on physical and microphysical levels and the possible correlation of the latter with real material structure, we will stress their independent (invariant) character as parameters, stipulating the spatial-temporal structure of the solid's dynamic rupture process itself. Thus, the main physical meaning of the mentioned structural values is reduced to the postulate, that on the given scale level we accept them for determining macroparameters of quasibrittle fracture processes (in the sense of thermodynamics).

An analogous viewpoint on the interpretation of the fracture linear dimension was expressed in the works of R. V. Goldstein and N. M. Osipenko [5, 6]. An apparent formality of such an introduction of determining factors is the advantage of the quoted works, as it is not connected with a unique physical interpretation, and installs a 'bridge' between different interpretations of microphysical properties of the material (crystals, polycrystals, polymers, composites, rocks) and the macromechanics of their fracture. Much more important is to choose the determining parameters as measurable, according to the basic principles of physics, i. e. directly or indirectly determined from experiments. It is clear from the aforesaid, and will be shown later, that this will be possible for $d$ and $\tau$.

# CHAPTER 5
## 'INTACT' FRACTURE

We will consider materials without artificially made defects and concentrators, like cracks or sharp notches, to be 'intact' materials. Let us examine the specific features of these materials' fracture and the possible methods of its modeling. In this chapter the works [30, 31, 45, 47, 94, 95, 98, 116] are used.

## §1. Cleavage Fracture in Solids: Dynamic Strength of Materials

Historically the first attempts to analyse cleavage were associated with the application of the critical stress criterion

$$\sigma \le \sigma_{\mathrm{c}}. \tag{5.1}$$

As the experiments have shown, this criterion could not describe many significant streaks of cleavage fracture, expressed by strength/temporal dependence and fracture spatial distribution. We notice that in the case of fracture, caused by a short-term pulse of large amplitude, the critical stress criterion contradicts the law of momentum variation. Thus, by accepting the fact that fracture is initiated by rectangular profile waves with duration $t_0$, for a threshold force pulse, we obtain $U_* = \sigma_{\mathrm{c}} t_0$, which according to the decrease of $t_0$ could be made as small as desired. It follows that even infinitesimal force pulses, which are not able to change material particles' momentum significantly, can cause fracture.

A number of phenomena, observed during the experiments, e. g. the phenomenon of dynamic branch appearance and distant cavity zones, and the necessity of their treatment, has led to a temporal criterion ([34, 35])

$$\int\limits_0^{t_*} \sigma(t)\,\mathrm{d}t \le J_{\mathrm{c}}. \tag{5.2}$$

Integral fracture characteristic (5.2) allows a theoretical justification of many important cleavage effects. However, experiments and fracture graphical analysis indicate a considerable role of the structure in this process. It is clear that the account of fracture structural peculiarities permits us to obtain some new information about the temporal dependence of material strength, whose explanation and theoretical description were problems of current interest hitherto. Many modern studies, undertaken in the field of dynamic deformation and fracture of materials, are oriented to this. At the same time, complicated physical fracture theories, taking account of structure processes, are not always effective during the analysis of practical engineering problems. That is why an elaboration of approaches, explaining and describing dynamic fracture peculiarities with the help of simple mechanical principles, is expedient.

Let us examine the aforementioned structural–temporal criterion, with the help of stress-field pulse characteristics and structural peculiarities of the fracture process. During the analysis of 'intact' media fracture the criterion takes the following form

$$\int_{t-\tau}^{t} \sigma(t') \, dt' \leq \sigma_c \tau. \tag{5.3}$$

For definiteness we assume $\tau = d/c$ and consider the classical one-dimensional cleavage problem (see, e. g., [35]). The condition (5.3) differs from the temporal criterion (5.2) in the existence of a structural fracture characteristic $\tau$. In order to determine what this approach will give us, we will examine the reflection of a triangular pulse of compression loading from the free end of a semi-infinite bar. Axis $Ox$ is directed along the bar, which is located at $x > 0$. The incident pulse is written as

$$\sigma_- = -P \left( 1 - \frac{ct + x}{ct_0} \right) [H(ct + x) - H(ct + x - ct_0)].$$

Here $P$ is the pulse amplitude, $t_0$ is its period, and $H(t)$ is the Heaviside function. The increasing section is absent. The stress profile, reflected from the free end, will have the following form

$$\sigma_+ = +P \left( 1 - \frac{ct - x}{ct_0} \right) [H(ct - x) - H(ct - x - ct_0)].$$

The combined stress is expressed as $\sigma = \sigma_- + \sigma_+$. For the first time the tensile stress maximum occurs at the point $x_0 = ct_0/2$. By introducing dimensionless values $T = ct/d$, $T_0 = ct_0/d$, we obtain

$$\sigma|_{x=x_0} = F + G, \tag{5.4}$$

$$F = P \left( \frac{1}{2} - \frac{T}{T_0} \right) \left[ H \left( T + \frac{T_0}{2} \right) - H \left( T - \frac{T_0}{2} \right) \right],$$

$$G = P \left( \frac{3}{2} - \frac{T}{T_0} \right) \left[ H \left( T + \frac{T_0}{2} \right) - H \left( T - \frac{3T_0}{2} \right) \right].$$

The rupture amplitude $P_*$, minimal for the given period $t_0$, will be found from the condition

$$\max_t I = \sigma_c, \qquad I = \int_{T-1}^{T} \sigma(T')\,dT'. \tag{5.5}$$

It follows from (5.4) that the maximum of $I(T)$ attains in the integration interval $(T_0/2, T_0/2 + 1)$. Moreover, $\max_t I(t) = PT_0/2$, if $T_0 \leq 1$ and $\max_t I(T) = P(T_0 - 1/2)/T_0$, if $T_0 \geq 1$. Due to (5.5), it follows that

$$T_* = \begin{cases} 1/\left[4(1 - \sigma_c/P_*)\right] + 1, & 1 \leq P_*/\sigma_c \leq 2, \\ 1 + \sigma_c/P_*, & P_*/\sigma_c \geq 2, \end{cases} \tag{5.6}$$

where $T_* = ct_*/t$ is the normalised time before fracture, defined as the moment when the integral form attains its critical value (5.3). The appropriate curve is shown in Fig. 5.1.

## §2. Temporal Dependence of Strength

The obtained correspondence between the fracture time $t_*$ and the threshold amplitude $P_*$ is called the temporal strength dependence. It shows that the dynamic strength is not a material constant but depends on time before fracture ('life time' of the specimen). In terms of this dependence the critical stress criterion (5.1) and the temporal approach of Nikiphorovski–Shemyakin (5.2) are on 'different poles'. The critical stress criterion qualitatively describes a quasistatic fracture at long times caused by continuous wave pulses. Experiments have shown that in the case of short-term loading we can observe a weak dependence of fracture time on threshold amplitude with a certain asymptote. This effect is called the dynamic branch phenomenon of temporal strength dependence.

The dynamic branch phenomenon defies definition both in traditional strength theory and in the temporal criterion (5.2). We notice that (5.2) gives a similar dependence under short term loading, but it does not cover the case of quasistatic loading. Dynamic branch location and its link-up with a quasistatic one remain unsolved. Thus, the critical stress criterion and the temporal criterion (5.2) describe only limit ends of the temporal strength dependence. As was stated above an introduction of a structural element allows construction of a unified curve of temporal strength dependence (Fig. 5.1). Static and dynamic branches turn out to be connected by a smooth passage. The physical meaning of the horizontal asymptote is the following: in the accepted assumption ($\tau = d/c$) it corresponds to the transmission time of interaction energy between structure elements. So, for aluminum alloy B95: ($\sigma_c = 460\,\text{MPa}$; $K_{Ic} = 37\,\text{MPa}\sqrt{\text{m}}$; $c = 6500\,\text{m/s}$):

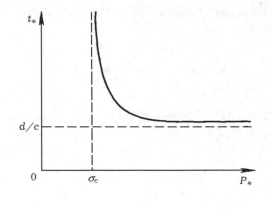

Fig. 5.1

$d/c = 2K_{\text{Ic}}^2/(\pi c\sigma_{\text{c}}^2) \approx 0.6\,\mu\text{s}$. It follows from the obtained formulas that the threshold amplitude of dynamic loading (cleavage strength) under modification of the loading duration at a range from 2 to $0.5\,\mu\text{s}$ increases from 600 to 1400 MPa. This fact agrees perfectly with the data of experiments from [9, 19]. The undertaken calculations have shown a satisfactory correspondence with the experiments, carried out for other materials.

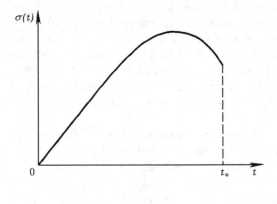

Fig. 5.2

As follows from (5.3)–(5.5), the fracture in a cleavage section happens with a delay, after a passage of the peak of local rupture stress (Fig. 5.2).

It is interesting to construct a fracture threshold, i. e. the dependence of minimal rupture pulse $U = PT_0/(2\sigma_0)$ on its duration. Such a threshold, obtained according to the critical stress criterion, is shown in Fig. 5.3 as a sloping dashed line. This line passes through the origin of the coordinate system; hence, in this case the fracture area in the plane $(T_0, U)$ adjoins the origin of coordinates. It denotes that even infinitesimal force pulses are able to cause fracture. This is absurd, because it is impossible to use the critical stress criterion for fracture analysis during short-term loading. The temporal criterion of Nikiphorovski–Shemyakin corrects this situation – the fracture threshold on the plane $(T_0, U)$ is traced by a horizontal line. It gives a finite threshold value for small durations; however, it does not link up with quasistatics at long times.

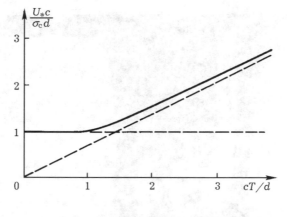

Fig. 5.3

The structural–temporal criterion (5.3) gives a unified threshold curve (full line in Fig. 5.3), suitable for the whole range of loading times. In our case the threshold curve is described by the following analytic formula

$$U = \begin{cases} 1, & T_0 \leq 1, \\ T_0^2/(2T_0 - 1), & T_0 \geq 1. \end{cases} \qquad (5.7)$$

In the limit of very long and very short pulses it corresponds respectively to the quasistatic and temporal criteria.

## §3. Fracture Zone Behavior Under Cleavage

Fracture zone behavior under cleavage is an extremely interesting subject to study. Classical approaches can not provide an adequate description. Thus, according to the critical stress criterion, the fracture zone can have a form of sequentially alternating cleavage sections. According to the temporal criterion fracture occurs continuously in the domain. This domain can have a finite extent, and is called a zone of continuous fractionation by V. S. Nikiphorovski and E. I. Shemyakin. The real fracture is characterised by both variants. In the K. B. Broberg work [63] the fracture domain has a form of fractionated parts, alternating with undisturbed bridges (Fig. 5.4 a). De facto, the zone of continuous subdivision is not continuous. It is also related to the destroyed domains in some other experiments (Fig. 5.4 b) (see, e. g., [113]). Moreover, as experiments show, the qualitative view of fracture domains depends on exterior pulse parameters, such as loading rate, amplitude and duration.

(a)

(b)

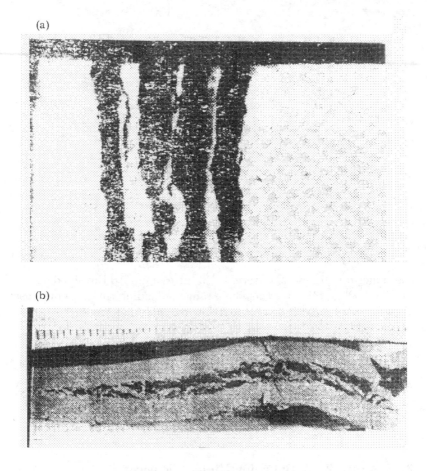

Fig. 5.4

Traditional approaches to the fracture zone study do not permit us to describe the whole variety of its geometry, obtained from the experiments. Thus, the use of the critical stress criterion (5.1) makes it possible to get a sequence of cleavage sections (cracks). The temporal criterion of

Nikiphorovski–Shemyakin makes it possible to predict a zone of continuous fractionation (see, e.g., [35]).

It is interesting that the structural–temporal criterion (4.8) permits modeling a fracture zone dynamics in a more complete way. Let us examine a scheme of calculation of destroyed domain parameters under the conditions of cleavage. We notice that in a one-dimensional situation of cleavage the rupture stresses are constant in the 'plane' of fracture, i.e. perpendicular to wave propagation. We suppose that the domain is linear, homogeneous and consists of successive identical structural strata with the thickness $b$. For definiteness, as a particular case, we could take $b = d$. An element (stratum) will be destroyed if the structural–temporal criterion in its middle has the following condition

$$\int\limits_{t-\tau}^{t} \sigma(t')\,\mathrm{d}t' \geq \sigma_{\mathrm{c}}\tau.$$

Let $t_*$ be the time when this condition is first fulfilled. Then, for $t < t_*$ the properties and the geometry are unchanged. At the temporal value $t = t_*$ the fracture of the whole structural stratum occurs. In this connection the whole part of the specimen, located between the face and the destroyed stratum, forms a cleavage plane. The destroyed stratum is an obstacle to the transmitted waves. Further reflection of the remaining waves, moving to the face and going from it, occurs from a new free surface.

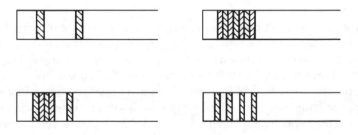

Fig. 5.5

Calculations, undertaken according to the aforementioned scheme, have shown that variations of fracture zones, shown in Fig. 5.5, can be realised only under a single trapezoidal pulse. The character of these zones could be significantly changed under the variation of rate of rise, amplitude, duration and the velocity of exterior loading decrease, which agrees completely with the existing picture of experimental studies.

## §4. On the Relationship of Quasistatic and
## Dynamic Mechanisms of Solid Fracture

The undertaken analysis permits a conclusion about the interconnection and range of developing of quasistatic and dynamic fracture mechanisms under cleavage. The main peculiarity of cleavage strength can be traced by means of the obtained diagram of temporal strength dependence. According to this diagram, the dynamic branch values, which are determined by structural characteristic $d/c$, correspond to a dynamic fracture mechanism. Moreover, the dynamic branch location does not correlate with the static strength of the material $\sigma_c$, which is confirmed by experiments. A transition zone, extending for times of the order of several structural intervals, reflects the joint manifestation of dynamic and quasistatic fracture mechanisms. In the examined situation both the dynamic fracture parameter and the critical power characteristic influence fracture threshold essentially. Significantly large fracture times, e. g. one order of magnitude larger than the time it takes for a wave to travel through the structure, can be examined as times corresponding to the action range of the quasistatic fracture mechanism. Such a fracture can be analysed with the help of the critical static stress criterion. Estimation and comparison with experience of fracture times for some materials (see, e.g., [9, 19]) lead to the conclusion that the range of essential influence of structural–temporal fracture singularities is determined by times of the order of several microseconds.

As an example, we will cite some calculated results of dynamic strength of rail steels RS 700 and RS 1100. The input data for rail steels are well-known [91]: RS 700: $\sigma_c = 780\,\text{MPa}$, $K_{\text{Ic}} = 70\,\text{MPa}\sqrt{\text{m}}$; RS 1100: $\sigma_c = 1160\,\text{MPa}$, $K_{\text{Ic}} = 48\,\text{MPa}\sqrt{\text{m}}$.

Steel RS 1100 was subjected to thermoprocessing (oil tempering from the temperature 920° and release at 540°) in order to create static rupture strength. The computed results of temporal strength dependence, for rail steels, are shown in Fig. 5.6. It is clear that steel RS 1100, in spite of a higher quasistatic rupture strength, has lesser strength under the conditions of high-rate shock loading, which is stipulated by its lower crack resistance.

The obtained conclusion is not trivial and demonstrates the necessity of a qualitative approach to constructional material selection with regard to corresponding velocity operating conditions. The structural–temporal approach allows the optimisation of this selection.

Fracture structural–temporal and force characteristics modification leads to a displacement of diagram parts of temporal strength dependence. Thus, the decrease of stress-wave propagation velocity changes the location of the dynamic branch in such a way that the material cleavage-strength increases. Therefore, heating of polymer material up to the temperature of a high-elastic state can lead to the increase of its cleavage strength. This conclusion agrees with experimental studies of cleavage-strength dependence of polymer com-

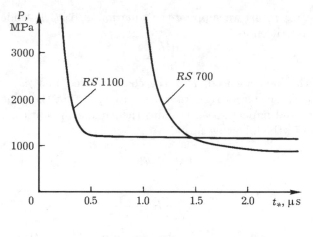

Fig. 5.6

posites on temperature ([4]). The threshold diagram in Fig. 5.3 permits the conclusion that fracture intensity depends on initial static rupture strength and on stress wave velocities. The latter is determined with the help of the elastic modulus and material density. With regard to this, we can conclude that more rigid and less massive materials can not qualitatively resist high-rate dynamic loading.

## §5. On Surface Erosion Under Hard Particle Impact

The dependence of the threshold pulse on its duration (Fig. 5.3), obtained with the help of the structural–temporal criterion, shows that, if we know the threshold values of extremely short loading pulses, we can determine an incubation time of fracture, corresponding to the given material. The latter allows the association of dynamic fracture and surface erosion phenomena in gas flows, containing hard particles. On the basis of fractographical analysis we can conclude [120] that the factor controlling erosion fracture is the formation of brittle annular cracks, produced by contact dynamic interaction of flying particles with the surface. Small particles with a radius of several dozens or hundreds of microns, used in the experiments on erosion fracture, produce extremely short rupture pulses during contact interaction with the surface. If we know their characteristics and the velocity value of threshold impact during which erosion fracture of a surface occurs, we can determine an elementary fracture 'quantum' and the corresponding incubation time.

Now we will show how the given scheme can be realised in the simplest approximation. Let a spherical hard particle with the radius $R$ fall on the surface of an elastic semi-space with velocity $v$. Following the classical Hertz scheme (see, e.g., [14]) we suppose the equation of particle (indenter) movement may be written as

$$m \frac{d^2 h}{dt^2} = -P, \tag{5.8}$$

where $h$ is the impact speed, $P$ is the contact force, and $m$ is the particle mass. In the classical approximation it is supposed that the relation between contact force and impact speed remains the same as in statics. This relation can be given in the following form

$$P(t) = kh^{3/2}, \tag{5.9}$$

where

$$k = \frac{4}{3} \sqrt{R} \frac{E}{(1 - \nu^2)}. \tag{5.10}$$

At the initial moment $dh/dt = v$; then, by integrating (5.8), we have

$$\frac{dh}{dt} = \sqrt{v^2 - \frac{4h^{5/2}}{5m}}. \tag{5.11}$$

The maximum approach $h_0$ is attained for $dh/dt = 0$; hence

$$h_0 = \left[ \frac{5mv^2}{4k} \right]^{2/5}. \tag{5.12}$$

To compute the impact duration we integrate (5.11) from the beginning of the interaction to the moment of maximum penetration

$$\int_0^{h_0} \frac{dh}{\sqrt{v^2 - 4kh^{5/2}/(5m)}} = \frac{t_0}{2},$$

where $t_0$ is the complete impact duration. Whence we have

$$t_0 = \frac{2h_0}{v} \int_0^1 \frac{d\gamma}{1 - \gamma^{5/2}} = 2.94 \frac{h_0}{v}. \tag{5.13}$$

Numerical integration permits the construction of the dependence of penetration as a function of time, i.e. via $h(t)$. This dependence is approximated with high precision by the expression ([14])

$$h(t) = h_0 \sin(2\pi/t_0). \tag{5.14}$$

The dependence of the maximum rupture stress on time at the surface, adjoining the contact platform, is computed according to the formula ([87])

$$\sigma(v, R, t) = \frac{1 - 2\nu}{2} \frac{P(t)}{\pi a^2(t)}, \tag{5.15}$$

where the radius of the contact spot $a(t)$ is determined as

$$a(t) = \left[ 3P(t)(1-\nu^2)\frac{R}{4E} \right]^{1/3},$$ (5.16)

and the contact force $P(t)$ is found with the help of (5.9)–(5.14).

Let $v$ be the threshold particle velocity, during which the material rupture happens. We introduce a function

$$f(v, R, \tau) = \max_t \int_{t-\tau}^{t} \sigma(v, R, s)\, ds - \sigma_c \tau.$$

In accordance with the structural–temporal criterion we determine an incubation time $\tau$ as a positive root of the equation

$$f(v, R, \tau) = 0,$$ (5.17)

for given values $v$ and $R$.

The obtained formulas can be used for calculation of the incubation time on the basis of experimental data on threshold velocity of surface erosion fracture.

Let aluminum alloy B95 with mechanical characteristics $E = 73\,\mathrm{GPa}$, $\nu = 0.3$, $\sigma = 456\,\mathrm{MPa}$ be subjected to erosion fracture with erodent characteristics $R = 150\,\mu\mathrm{m}$, $\rho = 2400\,\mathrm{kg}$ $(m = 3\pi\rho r R^3/4)$.

The dependence of incubation time $\tau$ on the threshold velocity of erosion fracture, calculated for the given parameters, is shown in Fig. 5.7. It is obvious that for a very large range of velocities, observed for aluminum alloys [57, 97], these methods produce adequate results.

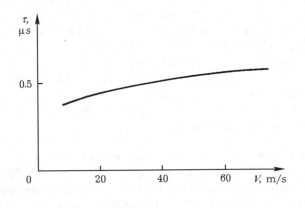

Fig. 5.7

The effective threshold particle velocity, at which the process of erosion surface fracture of the given material begins, must be determined experimentally and turns out to be equal to $v = 33\,\text{m/s}$ ([97]). Calculations according to the aforementioned formulas give the following values of characteristics of impact interaction between particles and surface: $t_0 = 0.29\,\mu\text{s}$, $h_0 = 3.46\,\mu\text{m}$. The study shows that function $f(v, R, \tau)$ has only one positive root (Fig. 5.8). The material incubation time, computed for the obtained data, turns out to be equal to $\tau = 0.5\,\mu\text{s}$.

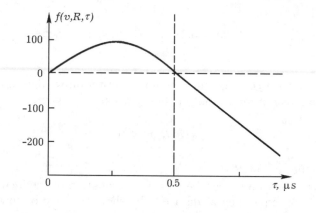

**Fig. 5.8**

The obtained value of incubation time permits us to construct a diagram of temporal dependence of strength for the indicated alloy. The corresponding computed curve, including not only static, but also dynamic branches is represented in Fig. 5.9. Experimental points, taken from the experiments on cleavage fracture for a given material ([8, 9]), included in the same picture, show the efficiency of the indicated methods of structural time evaluation on the basis of erosion data. It is noteworthy that approximately the same value for the structural time can be obtained by means of the simplified formula (4.15); $\sigma_c = 460\,\text{MPa}$; $K_{\text{Ic}} = 37\,\text{MPa}\sqrt{\text{m}}$; $c = 6500\,\text{m/s}$): $d/c = 2K_{\text{Ic}}^2/(pc\sigma_c^2) \approx 0.6\,\mu\text{s}$.

On the other hand, if we know the material incubation time, e. g. from experiments on cleavage fracture, we can determine the principal characteristics of the erosion process. The dependence of the erosion fracture threshold velocity of B95 alloy on the radius of erodent particles, calculated for $\tau = 0.5\,\mu\text{s}$, is represented in Fig. 5.10 (curve 1).

As these results show, the dependence is characterised by static and dynamic branches. The static part is characterised by a weak dependence of threshold velocity on the diameter of erodent particles. As opposed to that, the dynamic branch shows a rapid increase of threshold velocities with decreasing particle dimensions. Moreover, there is some characteristic length

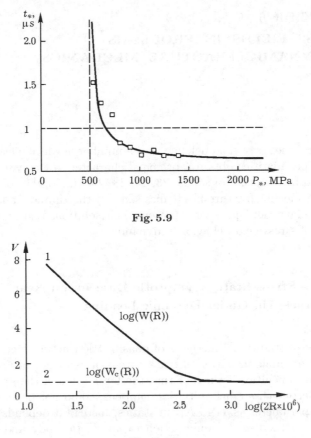

Fig. 5.9

Fig. 5.10

scale, in our case of the order of a few hundreds of microns, corresponding to the quick transition from the quasistatic regime to the dynamic one. The constructed theoretical curve qualitatively corresponds to well-known experimental observations ([51]). Notice that calculations, according to this scheme with the use of thetraditional critical stress criterion (Fig. 5.10, curve 2), can not explain the observed behavior of threshold velocities of erosion fracture.

CHAPTER 6

# STRESS FIELDS IN PROBLEMS
# OF DYNAMIC FRACTURE MECHANICS

In dynamics, the stress field behavior has a number of specific features. These important peculiarities obtain a principal character in the case of high-rate loading and fast rupture of solids [23, 24, 28, 29, 46, 50, 96]. In this chapter we will consider for fracture mechanics some of the simplest and traditional initial boundary value problems, present their solutions and notice some peculiarities of stress field behavior in dynamics.

## §1. On a Stress State Asymptotic Description Near
## a Crack Tip Under Dynamic Loading

Application of traditional methods of quasibrittle fracture mechanics to the problem of dynamic initiation of crack extension is rather problematic. One of the main difficulties is the fact that the limit value of the stress-intensity factor, under which a crack initiation happens, could not be supposed to be a material parameter as, e. g., $K_{Ic}$ in statics. Instead it depends both on the application method of dynamic action to an elastic body and on its loading history. This dependence is clearly stated in experiments on dynamic fracture initiation and is manifested in many ways, e. g. modification of the stress-intensity factor critical value under changes of amplitude, duration and velocity of the exterior action. Moreover, these effects occur most clearly in cases of the aforementioned rapid changes of conditions. Non-elasticity, non-linearity, temporal dependence of material properties and some other peculiarities of their behavior are often proposed as the main reasons causing these effects. In this case one often overlooks the fact that in an elastic problem the behavior of the dynamic stress field itself has a number of specific features, that could be reasons making the traditional description of fracture as a quasistatic one incorrect.

An asymptotic representation of the stress state near the crack tip is one of the most widely used mathematical tools for dynamic fracture analysis. Let us analyse the main physical principles and considerations, used to construct solutions.

We suppose that a plane elastic domain contains a crack (mathematical cut), and an impinging time-varying wave. Let us examine a spectral decomposition of the diffracted field. It is well known that for each harmonic $\omega$ the following asymptotic representation is valid

$$\sigma_{ij}(\omega, r, \theta) = \frac{K(\omega)}{\sqrt{2\pi r}} f_{ij}(\theta) + O(1), \quad r \to 0,$$

where $\sigma_{ij}$ is the stress tensor; and $r, \theta$ are polar coordinates at the crack tip. If we integrate (see, e. g., [60]) the reduced asymptotic expression with respect to the frequency $\omega$, we obtain the initial non-stationary solution near the crack tip in the following form

$$\sigma_{ij}(t, r, \theta) = \frac{K(t)}{\sqrt{2\pi r}} f_{ij}(\theta) + O(1), \quad r \to 0,$$

where $t$ is the time; $K(t)$, being the main characteristic of a quasibrittle fracture, is the stress-intensity factor.

This simple statement, often used to resolve concrete problems, contains a quite essential peculiarity. Namely, nothing guarantees the uniformity of the obtained asymptotic evaluation with respect to the time $t$ and to parameters characterising the geometry of the impinging wave. A detailed analysis permits us to observe that, under certain conditions, the dominating term of the asymptotic expansion is not 'the principal member of the solution'. Hence, in these cases the use of the stress-intensity factor for strength estimation does not seem possible.

The physical reason, generating this non-uniformity, is the presence of short waves in the spectral decomposition. Their integral ('batch') effect leads to a situation when the uniform convergence of the improper integral for times close to the moment of wave and crack interaction,

$$\int\limits_0^\infty \sigma_{ij}(\omega, r, \theta) e^{i\omega t} \, d\omega$$

vanishes. The use of the root asymptotically dominating term of solutions to the non-stationary problem for all temporal ranges is not justified under these conditions. Physically it is expressed, e. g., in the fact that the diffracted wave has on its front a feature, 'spoiling' the given asymptotic evaluation for small times.

## §2. Wave Stress Pulse Interaction with a Longitudinal Shear Crack

As an example, illustrating the proposals of the previous section, we will consider a problem of normal incidence of an antiplane step displacement wave at a semi-infinite crack $y = \pm 0$, $x \leq 0$, whose faces are free from exterior loads. At interior points of the plane, cut along $y = \pm 0$, $x \ll 0$, the stress field is described by the equations

$$\frac{\partial^2 u}{\partial x^2} + \frac{\partial^2 u}{\partial y^2} - \frac{1}{c^2}\frac{\partial^2 u}{\partial x^2} = 0,$$

$$\sigma_{xz} = \mu \frac{\partial u}{\partial x}, \quad \sigma_{yz} = \mu \frac{\partial u}{\partial y}. \tag{6.1}$$

The absence of fractions on the cut is expressed by the relation

$$\frac{\partial u}{\partial x} = 0, \qquad y = \pm 0, \quad x \le 0. \tag{6.2}$$

At times preceding the interaction of the impinging wave and the crack, the wave field is described by the function

$$u|_{t<0} = U H(ct + y), \quad y > 0, \tag{6.3}$$

where $H(x)$ is the Heaviside function; $U$ is the impact loading intensity, which, in this case, we suppose constant. Physically, it corresponds to a short-wave stress pulse, when the total force stress pulse is given by the expression

$$\Pi = \int \sigma_{yz} \, dt = \mu \frac{U}{c} \int \delta\left(t + \frac{y}{c}\right) dt = \frac{\mu U}{c}, \tag{6.4}$$

where $\delta(t)$ is the Dirac delta function. Let the condition

$$u = \text{const} + O\left(r^\beta\right), \qquad r = \left(x^2 + y^2\right)^{1/2} \to 0,$$

$$t \ge \varepsilon > 0, \quad \beta > 0, \tag{6.5}$$

also be fulfilled for an arbitrary positive number $\varepsilon$.

This condition means that the crack end is not an energy source ([52]), and ensures an unambiguous definition of the solution.

The solution of (6.1)–(6.5) can be found by S. L. Sobolev's method of functionally invariant solutions. However, in fracture mechanics, where the main problem is the determination of the stress-intensity factor, the method based on integral transformations and on the solution of the corresponding factorisation problem is more widely used. Using this method, we present the solution to (6.1)–(6.5) as a sum of impinging and reflected waves

$$u = U H(ct + x) + W(t, x, y), \qquad W \equiv 0, \qquad t < 0.$$

The function $W$ is determined as the following Fourier inversion

$$W(t, x, y) = \frac{1}{2\pi} \int\limits_{-\infty}^{+\infty} \varphi(\omega, x, y) e^{-i\omega ct} \, d\omega, \tag{6.6}$$

and $\varphi(\omega, x, y)$ is obtained from the solution of a related diffraction problem on a semi-infinite cut. Its solution can be obtained by the Wiener–Hopf factorisation method ([37]) and has a form

$$\varphi(\omega, x, y) =$$

$$= \mp \frac{U}{2\pi} \int\limits_{-\infty}^{+\infty} \frac{i \exp\left[\mp\omega \left(\gamma^2 - 1\right)^{1/2} y - i\omega\gamma x\right]}{\omega\gamma(\gamma - 1)^{1/2}} \, d\gamma, \quad \omega > 0. \tag{6.7}$$

The upper symbol corresponds to the points where $y > 0$, the lower to the points where $y < 0$; cuts in the Gaussian plane $\gamma$ are taken along the real axis in $\gamma < -1$ and $\gamma > 1$, and radical branches are fixed by the conditions

$$\mathrm{Im}(\gamma^2 - 1)^{1/2} < 0, \qquad -1 < \gamma < 1,$$

$$(\gamma + 1)^{1/2} > 0, \qquad \gamma > -1.$$

The contour of integration passes through the points $\gamma + 0i$ for $\gamma < -1$ and through the points $\gamma - 0i$ for $\gamma > -1$. The function $\varphi$ for $\omega < 0$ is determined by the relation $\varphi(\omega) = \overline{\varphi(-\omega)}$, the bar meaning complex conjugation.

Taking (6.5) into consideration when we search the entire complex function while solving the problem by the factorisation method, provides the possibility to reveal the corresponding behavior of the sought functions, corresponding to the Fourier transformation at infinity. This behavior provides the asymptotic stress state near the crack tip

$$\sigma_{xz} = -\frac{K(t)}{\sqrt{2\pi r}} \sin \frac{\theta}{2}, \quad \sigma_{yz} = -\frac{K(t)}{\sqrt{2\pi r}} \cos \frac{\theta}{2}, \tag{6.8}$$

$$r \to 0, \quad r \ll ct.$$

Whence it is clear that the stress-intensity factor $K(t) = \sqrt{2}U/(\pi ct)^{1/2}$ increases infinitely for $t \to 0$, and its expression can be obtained, as is commonly done by integration with respect to the stress-intensity factor frequency of the corresponding stationary problem.

Formal use of the critical stress-intensity factor criterion $K(t) \leq K_{\mathrm{Ic}}$ leads to a paradoxical conclusion: there is always such a time, when the criterion condition is accomplished. Hence, a fracture occurs for a total force stress pulse $\Pi$ (6.4) in the impinging wave. Griffith's energetical balance scheme for the solution of (6.8) gives the same result: energy flux, directed to the crack tip, tends to infinity for $t \to 0$, meaning that the given critical condition is satisfied for any value of $\Pi$.

Apart from the already mentioned physical reasons, the specified contradiction has also a pure-mathematical foundation.

Let us examine the exact solution. After substitution of (6.7) in (6.6) the double integral can be calculated by means of contour integration methods. Omitting intermediate computations, we give the final expressions for stress and displacement functions in the plane of the cut line

$$\sigma_{yz} = \mu U \left[ \sigma(ct) + H(ct - x) \frac{(ct)^{1/2}}{\pi ct(x-1)^{1/2}} \right],$$

$$y = 0, \quad x > 0,$$

$$\sigma_{xz} = 0, \quad u = U H(t), \quad y = 0, \quad x > 0,$$

$$\sigma_{xz} = \mp \frac{\mu H(ct + x)U}{\pi[-x(ct + x)]^{1/2}},$$

$$y = \pm 0, \quad x < 0,$$

$$u = U \left[ H(t) \pm H(t) \mp \frac{2}{\pi} H(ct + x) \arctan \left( \frac{ct + x}{-x} \right)^{1/2} \right],$$

$$y = \pm 0, \quad x < 0.$$

As follows from these expressions, for $t \leq r/c$ the stress behavior near the cut tip has a more complicated character than in (6.8), and on the front of the diffracted wave we can observe a root peculiarity. This means that the use of the stress-intensity factor in order to calculate a fracture possibility for $t \to 0$ could lead to erroneous conclusions. Thus, the traditional estimation, corresponding to asymptotics of square root singular stresses, is not uniform in dynamics. That is why, at times, close to the beginning of the interaction of an exterior pulse and a crack, the solution is not described by the principal member of the traditional asymptotic representation. Such a description is possible only after a certain time, necessary to establish the corresponding asymptotic 'regime'.

## §3. On the Stress State Near the Tip of a Rapidly Growing Crack

Experimental data (see, e. g., [105]) and numerical calculation results [89] show that a high-rate crack propagation can not be adequately described by means of traditional asymptotic formulas. Let us study an analytical example, clarifying this situation.

We retrace the behavior of a growing semi-infinite crack of longitudinal shear $y = 0$, $x \leq l(t)$, $l(0) = 0$. On its faces there is a uniformly distributed stress when the unbounded medium state is described by (6.1)

$$\sigma_{yz} = -p(t), \quad p(t) = PH(t), \quad y = 0, \quad x \leq l(t).$$

Let us suppose that for $t < 0$, all the points of the medium are in a quiescent state and at the cut tip the energy rate of change is finite. The solution of this problem is known (see, e.g., [15, 21])

$$\sigma_{yz}\Big|_{\substack{y=0 \\ x>l(t)}} = -\frac{1}{\pi\sqrt{x - l(t_*)}} \int_{x-ct}^{l(t_*)} p\left(t - \frac{x - s}{c}\right) \frac{\sqrt{l(t_*) - x}}{x - s} \, ds,$$

where $x - l(t_*) = c(t - t_*)$, i. e. $t_*$ has the meaning of time, when a signal, being received at the point $x$ at time $t$, propagates from the end of the growing cut. Whence, for the chosen loading, it follows that

$$\sigma_{yz}\Big|_{\substack{y=0 \\ x>l(t)}} = \frac{2P}{\pi}\left[\sqrt{\frac{ct - x + l(t_*)}{x - l(t_*)}} - \arctan\sqrt{\frac{ct - x + l(t_*)}{x - l(t_*)}}\right]. \quad (6.9)$$

Computing the asymptotic stress expression, we suppose that the crack velocity $v(t) = dl/dt$ is continuous. Then

$$l(t_*) - l(t) = v(t)(t_* - t) + o(t_* - t), \quad x \to l(t) + 0. \quad (6.10)$$

Inserting (6.10) in (6.9) and taking account of the equation for $t_*$, we have

$$\sigma_{yz}\Big|_{\substack{y=0 \\ x>l(t)}} = \frac{2P}{\pi}\left[\sqrt{\frac{t[c - v(t)] - x + l(t)}{x - l(t)}}\right.$$

$$\left. - \arctan\sqrt{\frac{t[c - v(t)] - x + l(t)}{x - l(t)}}\right] + o(1), \quad x \to l(t) + 0. \quad (6.11)$$

Essentially, (6.11) appears to be an asymptotic stress representation near the crack tip.

We suppose that

$$x - l(t) \ll t[c - v(t)]. \quad (6.12)$$

Then, it follows from (6.11) that

$$\sigma_{yz}\Big|_{\substack{y=0 \\ x>l(t)}} = \frac{2P}{\pi}\sqrt{\frac{t[c - v(t)]}{x - l(t)}} + o(1), \quad x \to l(t) + 0. \quad (6.13)$$

Thus, the transition from the exact solution to the principal member of root stress asymptotics (6.13) is correct only under the additional restriction (6.12), e. g. for sufficiently large times and velocities of crack growth, differing notably from the velocity of the shear wave. At the same time, an analysis based on the well-known fracture criteria and traditional root asymptotics can lead to mistakes when we deal with a high-rate material rupture. Thereby, some theoretical notions on the limit velocity of crack growth and on its behavior at close-to-limit energy flux velocity, found on the basis of traditional asymptotic formulas and the stress-intensity factor, require additional justification.

## §4. On the Influence of Regular Terms of the Stress-Field Asymptotic Representation

The studied examples show that the knowledge of the stress-field representation near the crack tip, obtained with the help of the principal (singular) term of the asymptotic expansion around the crack tip, generally accepted for static analyses, can be insufficient for analysis of dynamic fracture effects. It is natural, that the question of the influence of second-order terms of the expansion on computed results arises.

We will use a simple example to show that in many cases the concrete physical situation can be reflected more exactly with the help of a second term of the expanded asymptotic solution.

Let us consider an unbounded elastic plane with a rectilinear, semi-infinite crack $x \leq 0$, $y = 0$. The corresponding system of equations is given by

$$(\lambda + \mu) \operatorname{grad} \operatorname{div} \boldsymbol{w} + \mu \Delta \boldsymbol{w} = \rho \frac{\partial^2 \boldsymbol{w}}{\partial t^2}.$$

Here $\lambda$, $\mu$ are the Lamé constants; $\boldsymbol{w} = (u, v)$, where $u$, $v$ are components of the displacement vector.

Let a trapezoidal temporal profile stress be applied to the crack faces

$$\sigma_y = -V \left[ tH(t) - (t - t_0)H(t - t_0) \right],$$

where $t_0$ is the time of stress growth; $V$ is the loading rate. The scheme to solve this problem by the factorisation method is well known (see, e. g., [60]). The asymptotic expression of maximum tensile stress at the crack extension is given by

$$\sigma_y = \frac{K_{\mathrm{I}}(t)}{\sqrt{2\pi x}} - V \left[ tH(t) - (t - t_0)H(t - t_0) \right] + o(1), \qquad x \to 0, \qquad (6.14)$$

where

$$K_I(t) = V\varphi(c_1, c_2) \left[ t^{3/2} H(t) - (t - t_0)^{3/2} H(t - t_0) \right];$$

$$\varphi(c_1, c_2) = \frac{8c_2\sqrt{c_1^2 - c_2^2}}{3c_1\sqrt{\pi c_1}};$$

(6.15)

$c_1$, $c_2$ are the velocities of longitudinal and transverse waves. An asymptotic expression (6.14) is the approximate expression for the desired stress. The second term of this expression has the physical meaning of stresses applied to the crack faces.

Now we assume that under the previously given geometrical conditions fracture is initiated by a double-sided plane stress wave, incoming to the crack,

$$\sigma_y = \frac{V}{2} \left[ \left( t + \frac{y}{c} \right) H \left( t + \frac{y}{c} \right) - \left( t - t_0 + \frac{y}{c} \right) H \left( t - t_0 + \frac{y}{c} \right) \right.$$
$$\left. + \left( t - \frac{y}{c} \right) H \left( t - \frac{y}{c} \right) - \left( t - t_0 - \frac{y}{c} \right) H \left( t - t_0 - \frac{y}{c} \right) \right].$$

The crack faces remain free from fractions: $\sigma_{xy} = 0$, $\sigma_y = 0$. Then, the asymptotic expression of the maximum tensile stress on the crack extension is given by

$$\sigma_y = \frac{K_I(t)}{\sqrt{2\pi x}} + o(1), \qquad x \to 0,$$

where the stress-intensity factor is still determined by (6.15). In comparison with the previous case, in this situation there are larger stresses near the crack tip. This is stipulated by the existence of a transmitted wave. Thus, the current stress-intensity factor value is the same in both cases, and the difference in the mode of loading gives information on the second term of the asymptotic representation of the solution.

This difference could be essential for tests under high-rate loading. As follows from experiments, the fracture is faster when the loading rate increases. The stress-intensity factor value increases rather slowly with the change of time, i. e. as $t^\alpha$ ($\alpha > 0$). Therefore, for small fracture times, the process of the material rupture is 'controlled' by both singular and regular members of the asymptotic expansion.

Ignoring of the last circumstance leads to contradictions in interpretation of experimental data on fracture initiation. So, K. Ravi-Chandar and W. G. Knauss [104] got essentially higher values of the dynamic fracture viscosity for Homalite-100, than J. W. Dally and D. B. Barker [67], who had been conducting tests by modeling the same temporal dependencies for current stress-intensity factor values. The first author were using a caustic method to determine the named intensity factor, while the latter used the photo-elasticity method in combination with strain-gage measurements.

In the discussion of the accuracy and correctness of the methods used, which arose between the authors, the aforementioned difference in the modes of dynamic effect creation was not taken into account, they were just observed in two given sets of experiments: in the first case the stress was applied directly to the crack faces, in the second case the fracture was initiated by a double-sided plane tensile wave, incoming to the crack.

Further on it will be shown that an account of this circumstance, by means of the structural–temporal criterion of incubation time, permits a theoretical interpretation of such 'controversial' situations.

# CHAPTER 7

# DYNAMIC FRACTURE NEAR THE CRACK TIP

It is well known that when formulating the macrorupture criterion, complementing the solid-medium mechanics equations, one has to take into account the most important peculiarity of dynamic fracture – the existence of not only a spatial but also a temporal structure of the process. This circumstance must be reflected while choosing criterion-determining parameters and test methods of dynamic strength properties of a material. Structural–temporal criteria, already considered in the previous chapters, permit taking this dynamic fracture peculiarity into account and modeling the process of crack-growth initiation under the action of impact pulses.

In this chapter we examine some principal peculiarities, and present calculation methods and an interpretation of the well-known high-rate fracture effects of elastic bodies with cracks [23–29, 31, 49, 99].

## §1. Threshold Pulses of Impact Loading

An essential contribution to the solution of the problem of taking the temporal structure of the dynamic fracture process into account comes with the introduction of the already mentioned incubation time concept, which was suggested and developed by J. F. Kalthoff and D. A. Shockey [78], H. Homma et al. [73] and D. A. Shockey et al. [112]. Experiments, described in these works, testify that in the case of macrocrack growth initiation with the help of intensive short pulses, threshold amplitude values, obtained experimentally, turn out to be significantly greater than those stipulated by a traditional critical stress intensity factor criterion. That is why J. F. Kalthoff and D. A. Shockey [78] suggested that one should discard this criterion and accept the fact that fracture occurs when the current value of the dynamic stress-intensity factor $K_I(t)$ exceeds the value of the dynamic fracture viscosity $K_{Id}$ during some minimum time $t_{inc}$. The incubation time $t_{inc}$ is considered to be a material constant, connected with structural processes.

Experimental determination of the incubation time is accompanied by a very cumbersome procedure, requiring multiple specimen tests for different values of pulse action duration and complicated numerical calculations ([73, 112]). A priori learning of the functional dependence of the dynamic fracture viscosity on the history of loading is also essential for the minimum-time criterion.

In Chap. 4 we have examined another approach for fracture analysis, based on the structural–temporal criterion:

$$\frac{1}{\tau} \int_{t-\tau}^{t} ds \, \frac{1}{d} \int_{0}^{d} \sigma(s,r) \, dr \leq \sigma_{\mathrm{c}}, \tag{7.1}$$

where $\tau$ and $d$ are the structural time of fracture and its structural dimension; $\sigma_{\mathrm{c}}$ is the material static strength; and $\sigma(t,r)$ is the maximum tensile stress near the crack tip $(r = 0)$.

Structural dimension $d$ is determined according to data of quasistatic tests on cracked specimens. In the case of a generalised plane-strain condition it can be expressed by means of static fracture viscosity and strength by a simple formula [22]

$$d = \frac{2K_{\mathrm{Ic}}^2}{\pi \sigma_{\mathrm{c}}^2}.$$

According to this approach, $\sigma_{\mathrm{c}}$, $K_{\mathrm{Ic}}$ and $\tau$ form a system of determining parameters, describing material strength properties. The structural fracture time $\tau$ is responsible for dynamic peculiarities of brittle fracture and must be found experimentally for each material.

We will show that in experiments, carried out by J.F. Kalthoff and D.A. Shockey [78], H. Homma et al. [73] and D.A. Shockey et al. [112], the structural time $\tau$ may be interpreted as $t_{\mathrm{inc}}$.

Let an infinite plate have a crack $x \leq 0$, $y = 0$, and an incident rectangular profile stress wave

$$\sigma_y = P\left[H\left(t + \frac{y}{c}\right) - H\left(t + \frac{y}{c} - T\right)\right], \quad \sigma_{xy} = 0, \quad t < 0, \tag{7.2}$$

where $H(t)$ is the Heaviside function. We will find, for the given duration $T$, the minimum amplitude of an external pulse that will initiate crack growth. The asymptotic expression of the maximum tensile stress, that corresponds to the impulse (7.2), on the crack extension for $t > 0$ has the following form

$$\sigma_y = \frac{K_{\mathrm{I}}(t)}{\sqrt{2\pi r}} + o(1), \quad r \to 0,$$

$$K_{\mathrm{I}}(t) = P\varphi(c_1, c_2)f(t), \quad \varphi(c_1, c_2) = \frac{4c_2\sqrt{c_1^2 - c_2^2}}{c_1\sqrt{\pi c_1}}, \tag{7.3}$$

$$f(t) = \sqrt{t}\,H(t) - \sqrt{t-T}\,H(t-T),$$

where $c_1$, $c_2$ are speeds of the longitudinal and the transverse waves. According to (7.1) and (7.3) the expression of minimum amplitude, that leads to fracture, obtains the following form

$$P_1 = \frac{\tau K_{\mathrm{Ic}}}{\varphi(c_1, c_2)\max\limits_{t} \int\limits_{t-\tau}^{t} f(s)\, ds}. \tag{7.4}$$

At the same time, from the traditional critical stress-intensity factor criterion, it follows that the minimum stress amplitude is given by

$$P_2 = \frac{K_{\text{Ic}}}{\varphi(c_1, c_2) \max\limits_{t} f(t)}. \tag{7.5}$$

Now, we note that

$$\max\limits_{t} \frac{1}{\tau} \int\limits_{t-\tau}^{t} f(s)\, \mathrm{d}s < \max\limits_{t} f(t),$$

whence it follows that $P_1 > P_2$, whict is reflected in the abovementioned experiments of J. F. Kalthoff and D. A. Shockey [78], H. Homma et al. [73] and D. A. Shockey et al. [112]. They stated that for long cracks (short pulses) the values of the minimum breaking amplitude turn out to be greater than those obtained according to the traditional stress-intensity factor criterion.

The temporal dependence of the stress-intensity factor criterion is represented in Fig. 7.1.

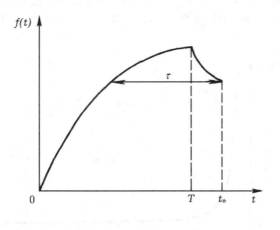

**Fig. 7.1**

As calculations under (7.1) show, the initiation of crack growth happens with a delay, i.e. during the decrease stage of local stress-field intensity near the tip. At the fracture moment $t_*$ the integral $\int_{t-\tau}^{t} f(s)\, \mathrm{d}s$ has its maximum value, hence $f(t_* - \tau) = f(t_*)$. According to the monotonicity of the function $f(t)$, we obtain, that in the analysed case $\tau$ is the time during which the stress-intensity factor exceeds the value $K_{\text{Id}} = K_{\text{I}}(t_*)$.

We also notice that computed values of dynamic viscosity under the examined conditions turn out to be inferior to the corresponding quasistatic value

$$K_{\mathrm{Id}} = P_1 \varphi(c_1, c_2) f(t_*) = \frac{\tau K_{\mathrm{Ic}} f(t_*)}{\int\limits_{t_*-\tau}^{t_*} f(s)\,ds} < K_{\mathrm{Ic}}, \qquad (7.6)$$

as is observed in experiments [73, 78, 112].

This reasoning remains true for a reasonable arbitrary temporal profile of a single pulse, providing a monotonic increase followed by a decrease of stress intensities.

So, the analysis of fracture caused by threshold pulses allows the observation that structural parameter $\tau$ has all the formal properties of an incubation time from the minimum-time criterion, and the problem of initiation of macrocrack growth it can be taken that

$$\tau = t_{\mathrm{inc}}. \qquad (7.7)$$

Criterion (7.1) and (7.7) permit efficient calculation of the values of external loading parameters.

Let us take an average incubation time, found in experiments on fracture of metal plates with a macrocrack: $t_{\mathrm{inc}} \approx 10\,\mu\mathrm{s}$. Calculating the threshold amplitude values in accordance with (7.4)–(7.7) we obtain values for the relative difference $Q = [(P_1 - P_2)/P_2] \times 100\,\%$ thoroughly in line with the data of experimental observations from [73, 112]. The corresponding dependence is presented in Fig. 7.2.

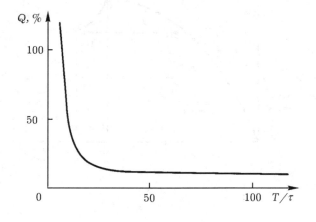

**Fig. 7.2**

Formulas (7.4) and (7.7) allow determination of the incubation time of a material fracture at the known threshold value of external pulse amplitude. Thus, for steel 4340 from [73] we have experimentally obtained a threshold amplitude value, within the region of 140–150 MPa, for $T \approx 20\,\mu\mathrm{s}$. According to (7.4) ($K_{\mathrm{Ic}} = 47\,\mathrm{MPa}\sqrt{\mathrm{m}}$, $c_1 = 6\,\mathrm{mm}/\mu\mathrm{s}$) we get $\tau \approx 7\,\mu\mathrm{s}$, coinciding with the incubation-time evaluation for this material from [73, 112].

For steel 4340, used in the experiments of H. Homma et al. [73], we have $c_1 = 6\,\text{mm}/\mu\text{s}$, $\nu = 0{,}3$, $K_{\text{Ic}} = 47\,\text{MPa}\sqrt{\text{m}}$, $\sigma_{\text{c}} = 1490\,\text{MPa}$, $t_{\text{inc}} = 7\,\mu\text{s}$. For $T = 18\,\mu\text{s}$ we get the critical value of amplitude $P_1 = 155\,\text{MPa}$ from (7.4) and (7.7). This value is in line with the experimental data of H. Homma et al. [73], when a similar critical value of external pulse amplitude, causing a crack 'jump' at the distance was calculated:

$$d = \frac{2K_{\text{Ic}}^2}{\pi\sigma_{\text{c}}^2} \approx 0.6\,mm.$$

## §2. On Loading-Rate Dependence of Dynamic Fracture Viscosity

Now, we suppose that there is a two-sided plane trapezoidal stress wave, incoming to the crack

$$\sigma_y = \frac{V}{2}\left[\left(t + \frac{y}{c}\right)H\left(t + \frac{y}{c}\right) - \left(t - t_0 + \frac{y}{c}\right)H\left(t - t_0 + \frac{y}{c}\right)\right.$$
$$\left. - \left(t - \frac{y}{c}\right)H\left(t - \frac{y}{c}\right) + \left(t - t_0 - \frac{y}{c}\right)H\left(t - t_0 - \frac{y}{c}\right)\right], \quad \sigma_{xy} = 0,$$

where $V = P/t_0$; $t_0$ is the given time of the applied stress increase up to the maximum value $P$. The corresponding asymptotic representation of the maximum normal stress for the crack extension is determined by (7.3), where

$$f(t) = \frac{2\left[t^{3/2}H(t) - (t - t_0)^{3/2}H(t - t_0)\right]}{3t_0}. \tag{7.8}$$

Let $t_*$ be the time before fracture, and $t_0$ be fixed. Using (7.1), (7.3) and (7.8) one can find the breaking amplitude $P_*$, appropriate for $t_*$. Then by calculating the critical stress-intensity factor value

$$K_{\text{Iq}} = K_{\text{I}}(t_*) = P_*\varphi(c_1, c_2)f(t_*),$$

we get

$$\frac{K_{\text{Iq}}}{K_{\text{Ic}}} = \frac{5}{2}\frac{\tilde{t}_*^{3/2} - \left(\tilde{t}_* - \tilde{t}_0\right)^{3/2}}{\tilde{t}_*^{5/2} - \left(\tilde{t}_* - 1\right)^{5/2} - \left(\tilde{t}_* - \tilde{t}_0\right)^{5/2} + \left(\tilde{t}_* - \tilde{t}_0 - 1\right)^{5/2}}, \tag{7.9}$$

where $\tilde{t}_* = t_*/\tau$; $\tilde{t}_0 = t_0/\tau$, and all power functions, for negative values of their arguments, are considered to be equal to zero. The corresponding

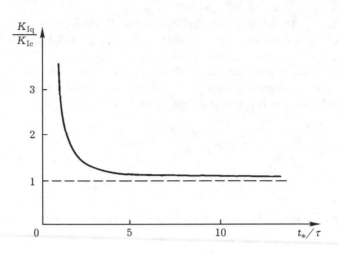

Fig. 7.3

graphical dependence is shown in Fig. 7.3. The same dependence was observed in many experiments (see, e. g., the survey [82]).

From (7.9) it follows that the link between the critical stress-intensity factor value and the fracture time depends on the time $t_0$ of the external loading increase. For bounded and semi-bounded domains this link depends also on geometric parameters of the problem. So, e. g., if $L$ is the crack length and $K_I(t) = PG(t, L)$, then, according to (7.1), we get

$$\frac{K_{Iq}}{K_{Ic}} = \frac{\tau G(t_*, L)}{\int\limits_{t_* - \tau}^{t_*} G(s, L)\, ds}.$$

Let us note that in (7.3), (7.8) and (7.9) the tension increasing ime can tend to zero, then.

$$\frac{K_{Iq}}{K_{Ic}} = \frac{3}{2}\frac{\tilde{t}_*^{1/2}}{\tilde{t}_*^{3/2} - (\tilde{t}_* - 1)^{3/2}},$$

which formally corresponds to an instantaneous application of constant stress. Thus, the qualitative link mode between $K_{Iq}$ and $t_*$ persists even under an 'infinite' loading rate.

We will show that the dynamic fracture viscosity can depend not only on the loading rate and the geometric parameters of the problem. Let us assume the fracture initiation caused by means of trapezoidal impact pulse action directly at the crack faces

$$\sigma_y = -V\left[tH(t) - (t - t_0)H(t - t_0)\right], \quad \sigma_{xy} = 0.$$

Then, in the infinitesimal order, we have on the crack plane

$$\sigma_y = \frac{K_I(t)}{\sqrt{2\pi r}} - V\left[tH(t) - (t - t_0)H(t - t_0)\right] + o(1), \quad r \to 0.$$

By the same reasoning as in the previous case, we get

$$\frac{K_{Iq}}{K_{Ic}} = \frac{5}{2} \frac{\tilde{t}_*^{3/2} - \left(\tilde{t}_* - \tilde{t}_0\right)^{3/2}}{\tilde{t}_*^{5/2} - \left(\tilde{t}_* - 1\right)^{5/2} - \left(\tilde{t}_* - \tilde{t}_0\right)^{5/2} + \left(\tilde{t}_* - \tilde{t}_0 - 1\right)^{5/2}}$$

$$\times \left\{1 + \frac{\tau V_*}{2\sigma_c}\left[\tilde{t}_*^2 - \left(\tilde{t}_* - 1\right)^2 - \left(\tilde{t}_* - \tilde{t}_0\right)^2 + \left(\tilde{t}_* - \tilde{t}_0 - 1\right)^2\right]\right\}. \quad (7.10)$$

For relatively great fracture times the critical stress-intensity factor value tends to the quasistatic value, and also

$$K_{Iq} = K_{Ic} + \frac{\lambda\tau}{t_*}, \qquad \frac{t_*}{\tau} \to \infty, \qquad \lambda = \text{const}.$$

Experiments, under conditions similar to the examined ones ([102]), have been carried out on specimens made from Homalite-100. A structural fracture time estimation for the named material can be made on the basis of comparison of data, found according to the theoretical formula (7.10), with experimental ones.

The estimated curve ($t_0 = 25\,\mu\text{s}$, $K_{Ic} = 0.48\,\text{MPa}\sqrt{\text{m}}$, $\tau = 8\,\mu\text{s}$) and the related experimental points, in logarithmic coordinates, are presented in Fig. 7.4.

K. Ravi-Chandar and W. G. Knauss have suggested an empirical formula

$$K_{Iq} = K_{Ic} + \frac{C}{t_*^2}, \qquad (7.11)$$

permitting an analytical description of experimental data. As follows from the results given above (see Fig. 7.4), (7.11) can be considered as an approximate power approach of the exact formula (7.10). In this case

$$C = \alpha\tau^2 K_{Ic}.$$

Formula (7.10), just as (7.9), demonstrates an increasing effect of the stress-intensity factor critical value with decreasing time before fracture, i.e. with the increase of loading rate. However, in comparison with the previous case, there are smaller stresses near the crack tip.

It should be noted that the current value of the stress-intensity factor in both cases is the same, and the difference in the way of loading manifests

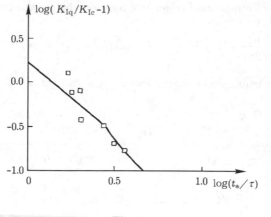

$\log(K_{Iq}/K_{Ic}-1)$

$\log(t_*/\tau)$

**Fig. 7.4**

itself in the value of the second term of the asymptotic representation of the solution. This is, ultimately, visualised via the experimentally obtained critical value of the stress-intensity factor of crack growth initiation.

As follows from (7.10) and (7.9), the dynamic fracture viscosity under wave loading turns out to be smaller than under the corresponding application of efforts directly to the crack faces. Value differences for $Q = \left[(K_{Iq}^{II} - K_{Iq}^{I})/K_{Iq}^{I}\right] \times 100\%$ for Homalite-100, where superscripts I and II correspond to the first and to the second cases respectively, are presented in Fig. 7.5. As computing results testify, the difference in the way of loading for longer times is hardly observable via the critical value the of stress-intensity factor.

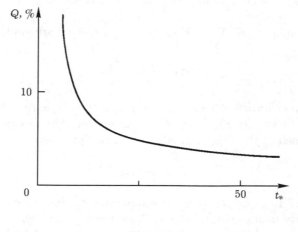

$Q, \%$

$t_*$

**Fig. 7.5**

At increasing loading rate, the fracture is faster but the dependence of dynamic fracture viscosity on the way of loading becomes more conspicuous. Evidently, the physical mechanism, causing such an effect, is an additional contribution of transmitted wave energy to the fracture, which is the more powerful the less the time is before fracture. It leads to a decrease of the stress-intensity factor critical value, which is necessary for creation of material rupture. This circumstance might be one of the reasons causing the apparent dispersion of results at experimental dynamic fracture viscosity determination. So, from the experiments on loading Homalite-100 with the help of high-intensive waves J. W. Dally and D. B. Barker [67] have obtained smaller critical values of the stress-intensity factor than K. Ravi-Chandar and W. G. Knauss [102]. This fact caused a discussion on the exactness of the experimental methods employed. The result, presented in Fig. 7.5, reflects the difference in dynamic fracture viscosity values, measured in the experiments mentioned.

## §3. Minimum and Maximum Pulses.
## Limit Characteristics of Material Dynamic Fracture

Results of the analysis process, carried out with regard to structural–temporal characteristics, reveals that dynamic effects depend on geometrical parameters, method and history of loading, and that their interpretation can not be reduced only to a velocity dependence of the dynamic fracture viscosity. It can be one of the explanations of great dispersion and inconsistency of experimental data on dynamic fracture viscosity of brittle materials.

Threshold pulses, studied in the beginning of the chapter, determine minimum, according to energy charges, loading conditions when crack initiation occurs. In this case the appearance of a fracture delay effect is essential: the criterion realisation takes place not at the initial stage of crack growth, but at the decrease of the current value of the stress intensity. This effect, that contradicts the classical mechanics of brittle fracture, is observed experimentally both in tests on cleavage (see, e. g., [9, 10, 35]), and in tests on fracture of cracked specimens (see, e. g., [73, 78, 112]). Here, the calculated critical values of the stress-intensity factor (dynamic fracture viscosity) turn out to be inferior to the corresponding quasistatic value $K_{Ic}$ for the given material. The latter is also a very important distinctive feature of the experiments mentioned.

Together with the examined threshold pulses, another situation was investigated, i. e. when the applied stress on the crack faces is maintained up to the moment of fracture. This guarantees a monotonic increase of the stress-intensity factor values during the whole structural–temporal interval $\tau$. Consequently, fulfilment of the inequality $K_{Iq} > K_{Ic}$ is also observed in corresponding tests [75, 82, 102].

Thus, the use of the structural–temporal criterion (7.1) for analysis of fast rupture in the crack-tip neighbourhood permits us to calculate the dynamic fracture viscosity of brittle materials. The results of the application of (7.1) to the problem of rectangular pulse action to the crack faces are presented in Fig. 7.6.

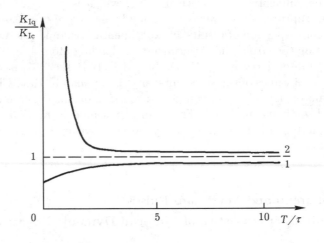

**Fig. 7.6**

Curve 1 in Fig. 7.6 determines the stress-intensity factor values at the moment of fracture under threshold pulses of duration $T$. In this case the material rupture occurs with a delay, i.e. at the stage of stress-intensity decrease near the crack tip, and $t_* > T$, where $t_*$ is the fracture time.

Curve 2 in Fig. 7.6 matches the case when a suddenly applied constant stress operates up to the fracture moment, so that $t_* = T$.

Such dispersion of dynamic fracture viscosity values, observed in the experiments, has become a reason for discussion on correctness and exactness of the experimental methods applied (see, e.g., [75, 85]).

The analysis suggests that the behavior of the critical stress-intensity factor is the principal peculiarity of dynamic fracture, stipulated by the discrete, structural–temporal nature of this process.

Let us analyse the behavior of fracture pulses under modification of their duration. Let the fracture near the crack tip be created by a rectangular-profile stress pulse (7.2). $U(T) = PT$ denotes the total force pulse of the external action. The computed dependence of the minimum fracture pulse $U = U_*(T)$ on its duration is presented in Fig. 7.7 (curve 1).

Note that if the applied pulse is inferior to $U_*$, but is of the same duration (through the amplitude decrease), fracture will not occur. Hence, all the points of the domain $UT$, situated below curve 1 (see Fig. 7.7), do not correspond to fracture. An important result is that the threhold pulse tends to a finite value for $T \to 0$. If we use the classical criterion of the critical stress-intensity factor, we can see (Fig. 7.7, dashed line), that fracture could be caused by even infinitesimal pulses, they only have to be sufficiently short,

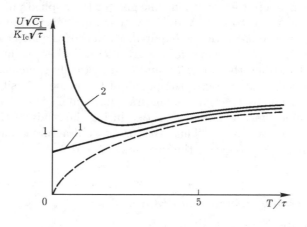

Fig. 7.7

which for certain is erroneous. Thus, under the action of short pulses it is advisable to use the classical criterion. However, the comparison of the computed threshold curves gives grounds to trust that for sufficiently great values of the duration $T$ ($\geq 10\tau$) it is possible to use successfully the classical criterion in order to evaluate fracture. Hence, the threshold pulse calculation reveals a fracture delay, the existence of its lower boundary in the coordinate plane $UT$ and the finiteness of fracture pulse values $U_*$ for short times of action.

Now, we fix the action time $T$, and let $U$ be the applied and $U_*$ the minimum rupture pulses. It is obvious that for $U > U_*$ fracture occurs, with the result that the fracture time exceeds the loading duration: $t_* > T$.

The question arises, how great the value of $U$ could be for the fixed $T$. Computed data show that if the pulse exceeds the threshold value to some degree, a fracture occurs with a smaller delay. The absence of delay corresponds to a coincidence of the fracture time with the pulse-action duration: $t_* = T$. Such a situation can be treated as an applied stress action up to the moment of fracture. In this case this occurs just when the stress-intensity factor reaches its maximum. An attempt of further pulse increasing with the help of an amplitude increase leads to a decrease of the duration of the applied stress action time, i. e. the fracture condition is fulfilled at time smaller than $T$.

Taking into consideration the aforesaid, we will name pulses, acting up to the moment of fracture, maximum rupture pulses and denote them by $U^*$. The dependence of the maximum rupture pulse on the loading time $U = U^*(T)$ is presented in Fig. 7.7 (curve 2).

Let us consider some peculiarities of fracture by means of maximum rupture pulses. At the decrease of action time one observes the increase of $U^*$ in comparison with $U_*$, and that $U^* \to \infty$ for $T \to 0$. So, if it is necessary to speed up the fracture the condition: 'the greater the applied pulse, the faster will fracture occur', is correct. And, lastly, to cause an instantaneous fracture an infinitely great pulse action is required. Evidently, the latter is connected with overcoming of medium inertia. For great $T$, i.e. when the medium 'manages' to start moving, the values $U_*$ and $U^*$ practically coincide.

As has already been noticed, the points of the plane $UT$, situated above curve 2, could not be reached: so, the fracture domain is under this curve, and at the same time, as has been stated during the study of the minimum rupture pulse, above curve 1. Thus, the 'probable' fracture on the plane $UT$ is the domain located between the two curves 1 and 2.

## §4. On Testing Principles of Material Dynamic Strength Properties

Let us classify some descriptive methods and construction material strength properties, and study their main possibilities ([93]) (Table 2).

### Table 2

| № | Method | Material parameters | Criterion |
|---|--------|---------------------|-----------|
| 1 | Classical approach of static fracture mechanics | $\sigma_c$, $K_{Ic}$ | $\sigma \leq \sigma_c$, $K \leq K_{Ic}$ |
| 2 | Classical approach of dynamic fracture mechanics | $\sigma_c^d(v)$, $K_{Ic}^d(v)$ | $\sigma(t) \leq \sigma_c^d$, $K(t) \leq K_{Ic}^d$ |
| 3 | Stanford J. F. Kalthoff, D. A. Shockey | $\sigma_c^d(v)$, $K_{Ic}^d(v)$, $t_{inc}$ | $\sigma(t) \leq \sigma_c^d$, Minimum-time criterion |
| 4 | Structural–temporal approach | $\sigma_c$, $K_{Ic}$, $\tau$ | Structural–temporal criterion |

In Table 2 the parameters $\sigma_c$ and $K_{Ic}$ are constants of the material, and $\sigma_c^d(v)$ and $K_{Ic}^d(v)$ are material functions, representing the dependence of critical characteristics on the loading rate $v$ [93].

The classical approach of dynamic fracture, based on principles of quasistatics and linear fracture mechanics, connects the material dynamic strength properties with two characteristics: $\sigma_c^d(v)$ and $K_{Ic}^d(v)$, which are to be considered as material functions. As has been noted, the direct transition of static boundary principles to dynamics problems turns out to be ineffectual: aside from the great experimental determination complexity of functions $\sigma_c^d(v)$ and $K_{Ic}^d(v)$ an investigator has to face a number of effects, that could not be explained by means of the given approach in principle. It occurs, e. g., that $\sigma_c^d(v)$ and $K_{Ic}^d(v)$ depend not only on loading rate, but on a whole number of other exterior action parameters. Such experimentally obtained values as $\sigma_c^d(v)$ and $K_{Ic}^d(v)$ are characterised by great dispersion, and, consequently, their behavior is poorly predictable.

The minimum-time criterion, elaborated by the Stanford International Research Center (California), includes a new material parameter $t_{inc}$, the incubation time. In comparison with the classical approach it has a number of new possibilities, in particular it allows the explanation of the fracture delay effect and the behavior of threshold pulses. An evident deficiency of this approach is its 'inheritance' of all the problems of the dynamic fracture classical approach: the analysis in compliance with the minimum-time criterion still requires a priori knowledge of velocity dependencies of material fracture strength and viscosity.

Fracture analysis from the point of view of the structural–temporal approach combines the evident advantages of the static fracture classical method and the efficiency of Stanford's approach. The determining fracture parameters are three material constants: $\sigma_c$, $K_{Ic}$ and $\tau$. The stress field dynamic intensity limit values, i. e. dynamic strength and fracture viscosity, can be considered as computed characteristics. According to the calculations their behavior is stipulated by a strong dependence on history and way of loading, which corresponds to the results of experiments. However, to determine the exterior loading parameter limit values we do not need an a priori knowledge of these dependencies. The established link between the fracture structural time $\tau$ and the incubation time $t_{inc}$ permits the use of well-known experimental methods [73, 78, 112] in order to obtain $\tau$ in the case of macrocrackns. An important peculiarity of the structural–temporal approach is that it allows evaluation of fracture near a macrocrack tip and fracture of 'intact' materials from a single position. Naturally, the fracture structural time for an 'intact' medium does not already coincide with macrocrack incubation time, as in this case, the material rupture occurs on another structural level. As follows from the analysis carried out, the structural time for 'intact' media can be determined on the basis of cleavage fracture experiments [8, 9, 20, 35]. Lastly, we observe that the structural–temporal criterion could be used to construct dependencies of exterior action critical characteristics, unique for statics and dynamics, as functions of these three material constants ($\sigma_c$, $K_{Ic}$ and $\tau$).

# CHAPTER 8

# DETERMINATION OF FRACTURE DIRECTION UNDER ASYMMETRIC-IMPACT ACTION

The problem, studied in this chapter, is connected wich the fracture 'mode change' effect. It was discovered experimentally by J. F. Kalthoff and S. Winkler [81], and then it was investigated by J. F. Kalthoff [76, 77], A. J. Rosakis et al. [107] and K. Ravi-Chandar [101].

## §1. Experimental Scheme and Fracture 'Mode Change' Effect

The scheme of the experiment is presented in Fig. 8.1.

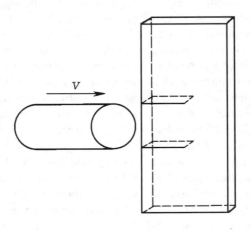

**Fig. 8.1**

A specimen with two parallel edge cracks is subjected to an impact of a cylinder shell, the diameter of which is equal to the distance between them. The shell initiates a compression wave, expanding in the domain between the mentioned cracks, creating, in its own turn, a specific asymmetric loading near its tip. It is supposed that in such a way the second mode of loading (dynamic transverse displacement) near the crack tip will be formed.

Crack behavior under such a loading turns out to be very complicated and unexpected. It can be reduced to three basic propositions:

1. Fracture does not happen during low rates of loading.
2. Fracture is observed when the velocity $V$ reaches a critical value: the edge crack extends catastrophically fast at an angle of approximately 70° to its initial direction (Fig. 8.2 a). This behavior is treated as a brittle fracture (tensile fracture failure mode), which can be explained from the classical brittle-fracture mechanics approaches, based on the principle of maximum tensile stress.

(a)

(b)

$V_0 < V < V_1$

$V > V_1$

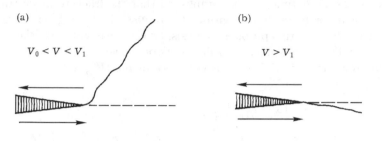

**Fig. 8.2**

3. A further impact velocity increase leads to the appearance of a new effect. It turns out that when $V$ reaches a certain critical value, a sharp change of fracture direction happens: the crack jumps a certain distance in the direction practically coinciding with the initial one (Fig. 8.2 b). At first sight, such a behavior is rather strange from the brittle-fracture mechanics point of view. Many investigators, including the originators of the discovery of this effect, connect it with the formation of local lines of sliding motion near the tip and consider it as a transition to a new shear (plastic) mode of dynamic fracture (shear band failure mode). The described effect was observed in experiments on a number of specimens of different materials (steel, vitreous polymers, polycarbonate, etc.).

Authors of the aforementioned works notice an absence of any theory explaining why the observed fracture mechanism changes (failure mode transition effect) when the impact velocity increases.

However, it is clear that the described situation must be analysed more thoroughly, first of all according to classical approaches. The problem is significantly complicated and has not been considered before as a classical fracture mechanics problem.

In this chapter an analysis of this problem is carried out. On the basis of the structural–temporal fracture criterion an interpretation of the observed 'transition' from the classical concept position of the opening mode will be given. The results given below have been obtained in [95]. Calculations were carried out for a hypothesised material, with properties similar to high-strength steel.

## §2. Asymmetrical-Impact Problem

Initial boundary-value problem analysis, modeling the described experimental conditions, has already been made by Y. J. Lee and L. B. Freund [88]. Supposing the material behavior to be linear elastic and the classical stress-intensity factor criterion to be true, they have obtained a good correspondence between theoretical and experimental results under low impact velocities. At the same time, it was established that the fracture mode transition effect could not be explained within the framework of the given model.

Let us continue this problem analysis, however we will now take into consideration both singular and regular parts of the solution.

We examine an initial boundary-value problem (Fig. 8.3)

$$u_x = \frac{\partial \varphi}{\partial x} + \frac{\partial \psi}{\partial y}, \qquad u_y = \frac{\partial \varphi}{\partial y} - \frac{\partial \psi}{\partial x},$$

$$\frac{\partial^2 \varphi}{\partial x^2} + \frac{\partial^2 \varphi}{\partial y^2} - a^2 \frac{\partial^2 \varphi}{\partial t^2} = 0, \qquad \frac{\partial^2 \psi}{\partial x^2} + \frac{\partial^2 \psi}{\partial y^2} - b^2 \frac{\partial^2 \psi}{\partial t^2} = 0,$$

$$a = \frac{1}{c_1} = \sqrt{\frac{\rho}{\lambda + 2\mu}}, \qquad b = \frac{1}{c_2} = \sqrt{\frac{\rho}{\mu}},$$

$$\sigma_x(-l, y, t) = 0, \quad \sigma_{xy}(-l, y, t) = 0, \quad y < 0,$$

$$u_x(-l, y, t) = \int_0^t v(t') \, dt', \quad \sigma_{xy}(-l, y, t) = 0, \quad y > 0,$$

$$\sigma_x(x, \pm 0, t) = 0, \quad \sigma_{xy}(x, \pm 0, t) = 0, \quad y < 0.$$

Here $\varphi$, $\psi$ are longitudinal and transverse wave potentials; $a$, $b$ are elastic wave reversal velocities respectively; $\lambda$, $\mu$ are The Lamé constants; $\rho$ is the density; and $v(t)$ is the loading rate, determined as $v(t) = VH(t)$, where $H(t)$ is the Heaviside function.

Boundary conditions are expressed per stress and displacement components.

The corresponding asymptotic solution for the stress component in the crack-tip neighborhood is given by the expression

$$\sigma_{ij} = \frac{K_I(t)}{\sqrt{2\pi r}} f_{ij}^{(I)}(\theta) + \frac{K_{II}(t)}{\sqrt{2\pi r}} f_{ij}^{(I)}(\theta)$$

$$+ \sigma_{ij}^{(R)}(r, \theta, t) + O(r^{1/2}), \quad r \to 0. \quad (8.1)$$

Here $r$, $\theta$ are polar coordinates at the crack tip.

The asymptotic solution, that we would like to use for analysis, consists of singular and regular parts. The fracture criterion, used in these cases, cannot only depend on the singular term. Therefore, we will use the structural–temporal criterion (7.1) to evaluate the crack-extension direction.

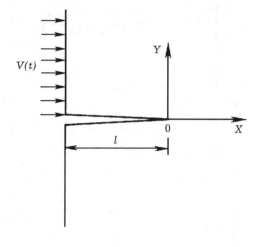

**Fig. 8.3**

The principal peculiarity of the dynamic fracture calculation according to this criterion is that it allows us to calculate the time before fracture. Anew and fairly simply, it permits determination of the fracture-extension direction in a cracked domain:

— we find the time before fracture $t_*$ for every ray, outgoing from the crack tip at angle $\theta$ $(-\pi/2 \leq \theta \leq \pi/2)$;

— we suppose the crack to be expanding to the direction, where the time before fracture $t_*$ is minimum.

The formulated aspects form the contents of the structural–temporal criterion under asymmetric loading, whose application to the corresponding dynamic stress field makes it possible to determine the exterior action critical parameters and the fracture-development direction.

## §3. Stress-Intensity Factors

The singular part of theasymptotic solution was studied by Y. J. Lee and L. B. Freund [88]. Let us analyse their results. Stress-intensity factors in the case of an impact pulse action $v(t) = \delta(t)$ are supposed to be represented by the following expressions

$$K_{\mathrm{I}}(t) = -\frac{\lambda'}{\sqrt{2\pi l}} \begin{cases} \dfrac{k}{\pi} \displaystyle\int_a^{h^*} \dfrac{2a^2(2h^2 - b^2)}{(h+a)\sqrt{(h-a)(h^*-h)}} P(h)\,dh, \\ \qquad\qquad\qquad\qquad a < h^* < b, \\[4pt] \dfrac{a\sqrt{2a}}{S_+(a)(c+a)\sqrt{h^*+h}} \\[4pt] \quad + \dfrac{a\sqrt{2a}}{S_-(c)\sqrt{c+h^*}(c^2 - a^2)}, \quad b < h^* < c, \\[4pt] \dfrac{a\sqrt{2a}}{S_+(a)(c+a)\sqrt{h^*+h}}, \qquad c < h^* < 3a, \end{cases} \tag{8.2}$$

$$K_{\mathrm{II}}(t) = -\frac{\lambda'}{\sqrt{2\pi l}} \begin{cases} \dfrac{k}{\pi} \displaystyle\int_a^{h^*} \dfrac{4a^2 h\sqrt{b-h}}{(h+a)\sqrt{(h^2-a^2)(h^*-h)}} P(h)\,dh, \\ \qquad\qquad\qquad\qquad a < h < b^*, \\[4pt] \dfrac{k}{\pi} \displaystyle\int_a^{b} \dfrac{4a^2 h\sqrt{b-h}}{(h+a)\sqrt{(h^2-a^2)(h^*-h)}} P(h)\,dh, \\ \qquad\qquad\qquad\qquad b < h < c^*, \\[4pt] \dfrac{k}{\pi} \displaystyle\int_a^{b} \dfrac{4a^2 h\sqrt{b-h}}{(h+a)\sqrt{(h^2-a^2)(h^*-h)}} P(h)\,dh \\[4pt] \quad + \dfrac{2a^2 c}{b^2(c^2-a^2)} \dfrac{2c^2 - b^2 - 2\sqrt{(c^2-a^2)(c^2-b^2)}}{S_-(c)\sqrt{c+b}\,\sqrt{h+h^*}}, \\ \qquad\qquad\qquad\qquad c < h^* < 3a. \end{cases} \tag{8.3}$$

Here $c = 1/c_R$ is a reverse velocity of the Rayleigh wave; $h = t^*/l$ and

$$P(h) = \frac{(h+c)(2h^2 - b^2)S_+(h)}{[(2h^2 - b^2)^4 + 16h^4(h^2 - a^2)(b^2 - h^2)]};$$

$$k = 2(b^2 - a^2);$$

$$S_\pm(\xi) = \exp\left\{ -\frac{1}{\pi} \int_a^b \arctan^{-1}\left[ \frac{4\eta^2 |\alpha\beta|}{(2\eta^2 - b^2)^2} \right] \frac{d\eta}{\eta \pm \xi} \right\};$$

$$\alpha(\xi) = (a^2 - \xi^2)^{1/2}; \quad \beta(\xi) = (b^2 - \xi^2)^{1/2}.$$

Stress-intensity factors for a loading velocity step function are obtained by the integration of (8.2) and (8.3). The corresponding temporal dependencies are presented in Fig. 8.4.

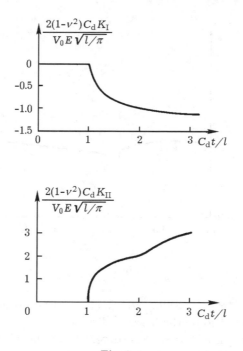

Fig. 8.4

One of the most important conclusions resulting from the analysis of stress-intensity factors is the fact that though the value $K_I$ is always inferior to $K_{II}$, the first mode value is, nevertheless, essential. Thus, the mixed mode of loading is, as a matter of fact, realised at the crack tip. The fracture-extension direction, under the circumstances, is controlled by the parameter ([111])

$$M_e = \frac{2}{\pi} \arctan\left(\frac{K_I}{K_{II}}\right).$$

The parameter $M_e$'s dependence on time in the particular case analysed is shown in Fig. 8.5. As follows from computing, its value changes inessentially within the temporal interval of special interest.

Supposing the crack expanding in a direction where the stress $\sigma_\theta$, computed on the basis of the asymptotic solution singular term, is maximum, we get the equation

$$3\sin\left(\frac{\theta}{2}\right) - \tan\left(\frac{\pi}{2}M_e\right)\sin\left(\frac{\theta}{2}\right)\cos\left(\frac{\theta}{2}\right) - 1 = 0. \tag{8.4}$$

The angle $\theta_*$, which corresponds to the maximum tensile stress $\sigma_\theta$ and is determined with the help of the extremum condition (8.4), depends on the time. This dependence is illustrated in Fig. 8.6. From this figure it is clear that, in the interval in which we are interested, the value $\theta_*$ changes in the

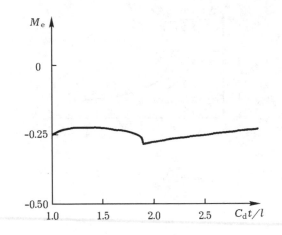

Fig. 8.5

neighborhood of $-78°$. According to the classical approach ([88]) this value should be considered to be the fracture direction.

The result, presented in Fig. 8.6, differs from the similar result, found by Y. J. Lee and L. B. Freund [88]. An error has been committed in their work [88]. The fracture direction $\theta_* \approx -63°$, found by these investigators, corresponds, in fact, to a point also satisfying the extremum condition (8.4), but being a local $\sigma_\theta$ stress minimum point.

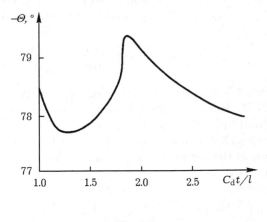

Fig. 8.6

We note that for a pure transverse displacement (mode II) the crack propagation direction, computed by means of a similar method, is equal to $\theta_* \approx -70°$. Thus, the existence of an additional contractive mode I leads to a deviation increase (and not to a decrease, as it was in [88]) of the crack-extension direction from its initial orientation.

The analysis, presented in this section, is based on the use of only the singular part of the asymptotic solution. The results found are in good correspondence with experimental data under low impact velocities. At the same time, it does not seem possible to explain the effect of crack-extension direction sharp change under impact velocity increase on the basis of the theory employed.

## §4. Fracture-Direction Change

We find the regular term $\sigma_{ij}^{(R)}(r, \theta, t)$, entering (8.1), as a compensating addition to the principal term of the asymptotic solution. The numerical studies have shown that waves, induced by an impact, create a tensile load near the boundary $x' > 0$, $y = 0$. Positive stresses, corresponding to these efforts, appear near the crack tip faster than a corresponding singular force field. Hence, during some temporal interval after a longitudinal wave arriving at the crack tip, an average tension, related to the solution regular part, exceeds, in absolute magnitude, the average tension of a singular wave field. It changes in principle the crack behavior under high impact velocities in comparison with the forecasting by results of only the singular part analysis of the asymptotic solution.

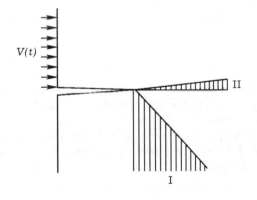

**Fig. 8.7**

Sectors near the crack tip, where the tensile loads act during certain time intervals, are presented in Fig. 8.7. The existence of such domains is revealed by means of calculations on the basis of (8.1), i.e. of the sum of singular and regular asymptotic solution parts. Analysis shows that both stress-intensity factors increase at initial times according to a power law ($\sim t^{1/2}$), while the finite wave field, corresponding to the solution regular part, appears instantly. Hence, fracture, occurring at initial times close to the moment of the crack tip loading, is determined both by singular and regular parts of the asymptotic solution. Such a situation is created under high impact velocities.

Computing the time before fracture $t_*$ for every direction ($\theta = $ const) near the crack tip with the help of the structural–temporal criterion, we get dependencies, presented in Fig. 8.8. Computed data permit the statement that there are no directions where the time before fracture would be finite under low impact velocities. When the impact velocity reaches a certain value, finite times to fracture are obtained in a continuous interval of directions. The minimum time before fracture in this interval is in the direction $\theta_* \approx -78°$.

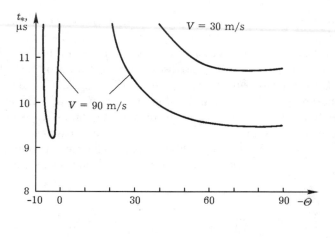

Fig. 8.8

As one can see from Fig. 8.8, there is a certain direction interval in the neighborhood of $\theta_*$, where the time before fracture hardly differs from the minimum one. It means that the real fracture direction in the neighborhood of $\theta_* \approx -78°$ could be characterised by a certain data scattering. Such a scatter was observed in the experiments accomplished by J. F. Kalthoff [76, 77, 81] and K. Ravi-Chandar [101].

The further increase of the impact velocity leads to the appearance of a second direction interval, where the time before fracture is finite. This is a narrow interval located in the neighborhood of angle $\theta_* \approx 4°$. If the impact velocity is sufficiently large, this value corresponds to the global time minimum before fracture for all ranges of directions. Thus, under high action velocities a crack 'jumps' in a direction almost equal to its initial orientation. This fracture duration should not be too long, because it is connected with wave fronts rapidly passing through the crack-tip neighborhood. The latter also characterises the experimentally observed 'fracture mode change'.

Thus, in both cases the fracture can be considered as elastic–brittle. The reason for particular crack behavior under an asymmetric-impact action is the forming of two possible extension zones (zones I and II) near the tip.

Under low impact velocities at the critical moment the average extension in the zone I exceeds the average extension in the zone II. Such a situation could be analysed on the basis of the singular part of the asymptotic solution, and the crack-extension direction could be predicted with the help of the classical approach ([88]).

Under high impact velocities the pattern changes: the tensile effort in zone II exceeds the tension in zone I. The transmitted wave possesses some rupture efforts in itself, which are impossible to neglect, and the crack-extension direction is controlled not only by the singular but also by the regular term of the asymptotic solution.

The structural–temporal fracture criterion permit an explanation of both cases from the brittle-fracture mechanics approach: the global time minimum before fracture can be located at different directions and depends on the intensity of the external impact action.

# CHAPTER 9
## ON MATERIALS YIELD MODELING
## UNDER HIGH-RATE LOADING[1]

In this chapter the incubation-time notion is used to construct criterion relations, describing the dynamic yield of materials. A similar approach was proposed by J. R. Klepaczko [122].

## §1. Experimental Studies of Material Yield
### Under High-Rate Loading

Let us examine a uniaxial loading of a metal specimen (tension or compression). In the case of quasistatic loading it is usually supposed that the material extends a into plastic state, if the applied stress attains a certain value, called the material-yield point; this value is an experimentally determined material constant. Thus, the static yield criterion can be written in the form of

$$\sigma = \sigma_0. \qquad (9.1)$$

In the high-rate loading case (9.1) loses its validity, moreover the dynamic properties of the material display themselves in two different ways, depending on the experimental mode. The first-mode experiments include those where the applied stress increases at a constant rate, i. e. $d\sigma/dt = $ const (see, e.g. [124]). The material becomes plastic when the applied stress attains values essentially exceeding (up to 2.5 times) the static yield point (see, e. g., [18]). The higher the loading rate is, the larger the values reached by the applied stress must be before yielding begins. By analogy with the static yield criterion (9.1) the given value has been called the dynamic yield point.

During the second-mode experiments the applied stress attains a certain value nearly instantaneously, and henceforth is retained constant up to the yield beginning. In spite of the fact that the applied stress exceeds this yield point significantly, there is a certain time before the yield begins; this time has been called the 'delay period', and the effect itself is called the 'yield delay'. The greater the applied stress is, the lesser the delay period ([56]).

Apart from the described direct material test methods, some indirect dynamic test methods are also used to study yielding. Their essence is to observe the dynamic loading construction. Analytical dependencies between the parameters, characterising dynamic material properties (usually the dynamic

---

[1]This chapter is written in collaboration with A. A. Gruzdkov.

yield point is chosen as such a parameter), and parameters that can be measured during the experiment, are established in terms of a simplified model process.

The common defect of indirect methods is the complexity when evaluating the influence of all assumptions, admitted in the theoretical model to obtain analytical dependencies, on the net result.

Among other approaches we also note dynamic ring dispensing ([64]), bar impact against a hard barrier, and transverse impact against a sheet ([3]).

## §2. Classical Description Methods of Material Dynamic Properties

First-mode material experiment peculiarities and the dynamic yield point are taken as the classical approach basis. It lies in the construction of experimental diagrams, with the help of direct or indirect study methods, connecting the dynamic yield point with the strain velocity, i. e. with such dependencies as

$$\sigma_{\text{yield}} = \sigma_{\text{yield}}(\dot{\varepsilon}). \tag{9.2}$$

An example of such an approach is a tabulation of dynamic factors corresponding to different loading rates. Taking this approach into account, we note its main imperfections.

Firstly, such a dependence as (9.2) does not account for loading-type influence, while in many problems the condition of loading-rate constancy is not realised even approximately. The stated approach is inapplicable, e. g., for material tests, where an impact loading is applied. Using the dynamic yield point, it is impossible to describe such effects as yield delay.

Secondly, we note an extreme experimental determination complexity of functional dependencies, a necessity to carry out a great number of experiments and a low reliability of the obtained diagrams.

The necessity to compare material testing results, realised with the help of different methods, and to study real problems, where the loading type may be very complicated, require the introduction of a criterion, considering loading history, i. e. an integral criterion. The construction complexity of experimental diagrams as well as the inconvenience of their application for calculations, testify that it is more advisable to describe the material dynamic behavior by a set of constants, having a concrete physical sense.

A criterion for mild steels, taking into account all the aforementioned requests, was suggested by J. D. Campbell [64]. The Campbell criterion can be set down as

$$\int_{0}^{t_{\text{yield}}} \left( \frac{\sigma(t)}{\sigma_0} \right)^{\alpha} \, \mathrm{d}t = \tau, \tag{9.3}$$

where $\alpha$ and $\tau$ are experimentally determined material constants.

This dynamic criterion was further developed in works of different investigators, where it was established that the constant on the right-hand side has a temporal dimension, and the static yield point can be used as $\sigma_0$. It has been noted that, though (9.3) is justified physically only for mild steels, quantitatively it describes well the behavior of other materials, including materials having no physical yield point. On the basis of experimental data the constants $\alpha$ and $\tau$ were calculated for different materials. It turned out that $\alpha$ changes in a range from 12 to 30, and $\tau$ can take values from several microseconds to one second.

Let us turn our attention to these constants in a physical sense. The yield dependence on loading rate is divided into two constituents. The constant $\tau$ is the material characteristic time. It is linked to temporal behavior parameters of certain processes inside the material structure and sets a temporal scale, determining, hereby, its relative capacity to display dynamic properties.

The numerical constant $\alpha$ describes in its turn the material behavior's absolute dependence on loading type. The greater is $\alpha$, the less are the displayed dynamic material properties. For materials with a large $\alpha$ the yield delay is not sensibly observed, and the yielding itself depends on the applied stress amplitude. In truth, (9.3) can be rewritten as

$$\left( \int_0^\xi \sigma^\alpha(\xi)\, \mathrm{d}\xi \right)^{1/\alpha} = \sigma_0,$$

where $\xi = t/\tau$ is non-dimensional time. The aforesaid becomes more clear if we take into account that

$$\left( \int_a^b f^\alpha(t)\, \mathrm{d}t \right)^{1/\alpha} \xrightarrow[\alpha \to +\infty]{} \max_{[a,b]} f(t).$$

The main imperfection of (9.3) is its coming into collision with the static yield criterion (9.1) for a long loading duration, because it confirms the fact that yielding can occur at an infinitesimally loading amplitude. The comparison of data, found according to (9.3), with the known experimental data (see, e.g., [64]) permits the observation that notable deviations appear in the case when the loading duration exceeds $\tau$. In this case, the yield stress still significantly exceeds the static yield point.

## §3. New Yield Criterion

To summarise, according to the aforesaid, we are in the following situation: under a low rate loading, it is necessary to use the static yield criterion (9.1), under a high rate loading, Campbell's criterion (9.3); the intermediate range is described neither by the first nor by the second criterion. Besides, the inconvenience of this situation is aggravated by the fact that the question whether the loading is of low rate or of high rate needs to be linked up with the material properties, i.e. the same loading could be of high rate for one material and of low rate for another. It makes the construction of a unified yielding criterion topical both for statics and for dynamics.

Let us study another approach, based on the incubation-time concept. In this case we proceed from the following premises:

1. There is a certain threshold stress value when the yielding can happen. If the applied stress does not attain this value, the material maintains its elastic state. According to the static yield criterion this threshold value is equal to the yield point $\sigma_0$.

2. As follows from experiments, greater delay periods correspond to smaller applied stress values.

3. From the two first premises, we can draw a conclusion about the existence of the greatest yield delay period as follows.

4. Data, received according to Campbell's criterion, are in good correspondence with experimental data for the loading duration not exceeding $\tau$. Using (9.3), we get the maximum possible delay period equal to $\tau$.

The greatest delay period corresponds to the maximum (incubation) time, while the material can 'resist' its transition into the plastic state under loading, capable, in principle, of causing yielding. On the basis of these premises we will accept a hypothesis, according to which a transition into a plastic state is influenced not by the whole loading history, but only by loading occurring in the period immediately before to the yield beginning. The duration of such a period is $\tau$.

Consequently, we come to the next yield criterion ([7][2] )

$$\frac{1}{\tau} \int\limits_{t_{\text{yield}}-\tau}^{t_{\text{yield}}} \left( \frac{\sigma(t)}{\sigma_0} \right)^\alpha \, dt = 1. \tag{9.4}$$

We suppose $\sigma(t) = 0$ for $t < 0$. Criterion (9.4), for external pulses with durations not exceeding $\tau$, coincides with (9.3). The dependence of the yield stress, computed according to (9.4), on loading duration in the case of $\sigma = $ const is in good correspondence with experimental data.

## §4. Constant-Rate Loading

The studied yield criterion is universal in the sense that it can be used for loadings of arbitrary form and duration. Let us analyse, as an example, a

---

[2]A similar approach was also proposed in [122].

constant-rate loading. The importance of this case is conditioned by two circumstances. Firstly, material tests under such loading are rather routine and can be used to compute constants, describing the dynamic material behavior ($\alpha$ and $\tau$). Besides, in many applied problems the substitution of the real stress velocity $\dot{\sigma}$, changing in time, by some value $\dot{\sigma}_{\text{mean}}$, averaged for all the process does not lead to a significant error.

We suppose $\dot{\sigma} = \text{const}$, so that

$$\sigma = \dot{\sigma} t H(t), \tag{9.5}$$

where $H(t)$ is the Heaviside function. We denote the yield beginning time $t_{\text{yield}}$, and the yield stress $\sigma_{\text{yield}}$. Then, from (9.5) we have $\sigma_{\text{yield}} = \dot{\sigma} t_{\text{yield}}$. We will construct the dependence of the yield stress on the loading rate. First, we consider the case of high-rate loading, i.e. the case $t_{\text{yield}} < \tau$. Then, the structural–temporal criterion (9.4) is just the same as Campbell's criterion (9.3).

Substituting (9.5) into (9.3), we obtain

$$(\dot{\sigma})^{\alpha} t_{\text{yield}}^{\alpha+1} = \tau \sigma_0^{\alpha}(\alpha + 1).$$

Excluding here $t_{\text{yield}}$ with the help of (9.5), we come to the formula

$$\sigma_{\text{yield}}^{\alpha+1} = \tau \dot{\sigma} \sigma_0^{\alpha}(\alpha + 1).$$

Multiplying and dividing the right-hand side of this equality by $\sigma_0$, we get

$$\sigma_{\text{yield}} = \sigma_0 (1 + \alpha)^{1/(\alpha+1)} \left( \frac{\dot{\sigma}}{\dot{\sigma}_0} \right)^{1/(\alpha+1)}, \tag{9.6}$$

where $\dot{\sigma}_0 = \sigma_0 / \tau$.

Changing the strain base velocity and taking into account that for a linear-elastic material $\dot{\sigma} = E\dot{\varepsilon}$, where $E$ is the Young modulus, (9.6) can be written in another form. Hence we obtain

$$\sigma_{\text{yield}} = \sigma_0 \left( \frac{\dot{\varepsilon}}{\dot{\varepsilon}_1} \right)^{1/(\alpha+1)}. \tag{9.7}$$

Here $\dot{\varepsilon} = \varepsilon / \tau(\alpha + 1)$, and $\varepsilon = \sigma/E$. We note that (9.7) remains true for $\dot{\varepsilon} \gg \dot{\varepsilon}_1$, though for strain velocities commensurable with $\dot{\varepsilon}_1$ it gives low estimated values of the yield stress. Therefore, the velocity $\dot{\varepsilon}_1$ has no certain physical sense, in contradiction to $\dot{\varepsilon}_0 = \dot{\sigma}_0/E$.

At first sight (9.7) seems to resemble the dependence, used in many works

$$\sigma_{\text{d}} = \sigma_{\text{st}} \left( \frac{\dot{\varepsilon}}{\dot{\varepsilon}_{\text{st}}} \right)^{\mu}. \tag{9.8}$$

However, we will establish their inherent difference. In (9.8) $\dot{\varepsilon}_{st}$ is the deformation velocity in quasistatic tests. Let us suppose (9.8) to be true for $\dot{\varepsilon} > \dot{\varepsilon}_{st}$, and for smaller velocities the static yield stress value can be introduced. Computing shows that $\dot{\varepsilon}_{st}$ is essentially inferior to $\dot{\varepsilon}_0$ and even to $\dot{\varepsilon}_1$.

While using (9.8), we ignore the existence of intermediate loading rates. High and intermediate rates are presented alike; naturally, this deteriorates the description exactness, particularly for very high loading rates. A similar opinion was also stated by J. D. Campbell [64].

We should note that (9.6) remains valid only for high loading rates $\dot{\sigma} > \dot{\sigma}_0$. For $\dot{\sigma} = \dot{\sigma}_0$ the yield stress notably exceeds the static yield point, and, consequently, the static yield criterion is inapplicable.

The dependence for intermediate rate loading, similar to (9.6), is found according to the common yield criterion (9.4). The dependence, common for the entire range of strain velocities between $t_{yield}$ and $\dot{\varepsilon}$, obtained with the help of (9.4), is written as

$$\left(\frac{t_{yield}}{\tau}\right)^{\alpha+1} - \left(\frac{t_{yield}}{\tau} - 1\right)^{\alpha+1} H\left(\frac{t_{yield}}{\tau} - 1\right) = \frac{(\alpha+1)\sigma_0^\alpha}{(\tau E\dot{\varepsilon})^\alpha}. \qquad (9.9)$$

An analogous dependence, resulting from the structural–temporal criterion of brittle fracture for a specimen from an 'intact' material (5.3) is presented as

$$\left(\frac{t_F}{\tau_c}\right)^2 - \left(\frac{t_F}{\tau_c} - 1\right)^2 H\left(\frac{t_F}{\tau_c} - 1\right) = \frac{2\sigma_c}{\tau_c E\dot{\varepsilon}}, \qquad (9.10)$$

where $t_F$ is the brittle fracture time, and $\tau_c$ is the structural fracture time.

Comparing (9.9) and (9.10), one can obtain curious conclusions on the possible fracture mode change.

We consider, as an example, a bar extension with the given strain velocity $\dot{\varepsilon}$. To describe the material behavior we use two models: the plastic yielding model and the fracture model. In first approximation it is natural to use a model, ignoring these two processes' interference, and to consider them as independent of each other. In this case to analyse the yielding and the fracture we use dynamic criteria (9.4) and (5.3), whence we get (9.9) and (9.10). Based on these equations we can calculate the resistance to plastic yielding ($\sigma_{yield} = E\dot{\varepsilon}t_{yield}$) and to fracture ($\sigma_F = E\dot{\varepsilon}t_F$), corresponding to different strain velocities. The case when the yielding stress is smaller than the fracture stress corresponds to a viscous fracture, and if the yield stress appears to be superior to the fracture stress, it corresponds to a brittle fracture. Such an approach was used by J. B. Friedman [58]; it is thoroughly efficacious for a qualitative description of a limiting plastic strain behavior.

An interesting point of the diagrams corresponds to the critical strain velocity, when the fracture mode changes. A strain velocity increase contributes to the transition from a viscous fracture to a brittle one.

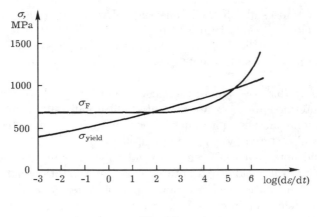

**Fig. 9.1**

The dependencies are presented in Fig. 9.1 for yield and fracture limit stresses, computed according to (9.9) and (9.4) for a hypothetical doped steel with characteristics $\sigma_0 = 400\,\text{NPa}$, $\sigma_c = 700\,\text{MPa}$, $\tau_c = 10^{-6}\,\text{s}$, $\tau = 0.1\,\text{s}$, $\alpha = 15$. They confirm the fact of the fracture mode change with the strain-velocity increase. Besides, the accepted model allows the observation that under very high loading rates there must be one more intersection point in the diagrams, and consequently, the brittle fracture must be replaced with the viscous fracture, if, of course, this model is still applicable for such high loading rates. It is to be noticed that there are experimental data confirming indirectly the existence of such a point.

# REFERENCES

1. *Alexandrov A.J. and Akhmetzyanov M.H.* Photoelastic Methods of Deformable Body Mechanics (Nauka, Moscow, 1973, 576 pp.). (In Russian.)

2. *Bolotin V.V.* Nonconservative Problems of Elastic Resistance Theory (Nauka, Moscow, 1961, 339 pp.). (In Russian.)

3. *Voloshenko-Klimovitskii Y.J.* Dynamic Yielding Limit (Nauka, Moscow, 1967, 179 pp.). (In Russian.)

4. *Golubev V.K., Novikov S.A., Sobolev Y.S. and Yukina N.A.* On heating influence on cleavage fracture of some polymer composites. Appl. Mech. and Thec. Phys. **28** (6), 140–145 (1987). (In Russian.)

5. *Goldstein R.V. and Osipenko N.M.* Fracture and Structure Formation. Reports Science Acad. USSR **240** (4), 829–832 (1978). (In Russian.)

6. *Goldstein R.V. and Osipenko N.M.* Structures and processes of formation fracture. Modeling of Seismic Process Development and Earthquakes Prognosis, Issue 1 (Nauka, Moscow, 1993, pp. 21–37.). (In Russian.)

7. *Gruzdkov A.A. and Petrov Y.V.* On metal yielding unique criterion under low and high-rate loading. Proceedings 1st All-Union Conference 'Technological Problems of Bearing Constructions Strength', Zaporozhje, 1991. Vol. 1, Part 2. pp. 287–293. (In Russian.)

8. *Zlatin N.A., Pugachev G.S., Mochalov S.M. and Bragov A.M.* Fracture process temporal regularities under intensive loading. Physics Solids **16** (6), 1752–1755 (1974). (In Russian.)

9. *Zlatin N.A., Pugachev G.S., Mochalov S.M. and Bragov A.M.* Temporal dependencies of metal strength at microsecond range durability. Physics Solids **17** (9), 2599–2602 (1975). (In Russian.)

10. *Zlatin N.A., Peschanskaya N.N. and Pugachev G.S.* On brittle solid delay fracture. Journal Tech. Phys. **56** (2), 403–406 (1986). (In Russian.)

11. *Kannel G.I. and Fortov V.E.* Condensed media mechanical properties under intensive pulse actions. Mechanics Successes **10** (3), 3–82 (1987). (In Russian.)

12. *Kachanov L.M.* Fundamentals of Fracture Mechanics (Nauka, Moscow, 1974, 312 pp.). (In Russian.)

13. *Kikoin A.K. and Kikoin I.K.* Molecular Physics (Nauka, Moscow, 1976, 500 pp.). (In Russian.)

14. *Kolesnikov Y.V. and Morozov E.M.* Contact Fracture Mechanics (Nauka, Moscow, 1989, 219 pp.). (In Russian.)

15. *Kostrov B.V., Nikitin L.V. and Flitman L.M.* Brittle fracture mechanics. News Science Acad. USSR. Solid Mechanics. 1969. N 3. P. 112–125. (In Russian.)

16. *Lavrentiev M.A. and Ishlinsky A.Y.* Dynamic forms of elastic system buckling. Reports Science Acad. USSR. **64** (6), 779–782 (1949). (In Russian.)

17. *Mazja V.G. and Nazarov C.A., Plamenevskii B.A.* Elliptic Boundary Problems Solution Asymptotics Under Singular Domain Perturbations (Tbilissi University Press, Tbilissi, 1981. 206 pp.). (In Russian.)

18. *Maiboroda V.P., Kravchuk A.S. and Hohlin N.M.* High-speed Deformation of Engineering Materials (Nauka, Moscow, 1986, 260 pp.). (In Russian.)

19. *Meshcheryakov Y.I.* Statistic model of cleavage surface formation and fracture criterion. Surface Physics, Chemistry, Mechanics. 1988. N 3. P. 101–111. (In Russian.)

20. *Meshcheryakov Y.I., Divakov A.K. and Kudryashov V.G.* On dynamic strength under cleavage and disruption. Physics Burning and Explosion **24** (2), 126–134 (1988). (In Russian.)

21. *Molchanov A.E. and Nikitin L.V.* Longitudinal shear crack dynamics after buckling. News Science. Acad. USSR. Solid Mechanics. 1972. N 2. P. 60–68. (In Russian.)

22. *Morozov N.F.* Mathematical Problems of Crack Theory (Nauka, Moscow, 1954, 255 pp.). (In Russian.)

23. *Morozov N.F. and Petrov Y.V.* Dynamic fracture viscosity in crack growth initiation problems. News Science Acad. USSR. Solid Mechanics. 1990. N 6. P. 108–111. (In Russian.)

24. *Morozov N.F. and Petrov Y.V.* On fracture analysis near fast growing crack tip. Vestnik Leningrad State University Series 1, Issue 1, 121–122 (1991). (In Russian.)

25. *Morozov N.F. and Petrov Y.V.* On structural time concept in dynamic fracture theory of brittle materials. Reports Russian Science Acad. **324** (5), 964–967 (1992). (In Russian.)

26. *Morozov N.F. and Petrov Y.V.* On structure–temporal description of high-speed dependence of dynamic brittle material fracture viscosity. News Russian Science Acad. Solid Mechanics. 1993. N 6. P. 100–104. (In Russian.)

27. *Morozov N.F. and Petrov Y.V.* On threshold velocities of solid surface erosive fracture. News Russian Science Acad. Solid Mechanics. 1996. N 3. P. 72–75. (In Russian.)

28. *Morozov N.F., Petrov Y.V. and Utkin A.A.* On fracture near crack tip under impact loading. Physical–Chemical Material Mechanics. 1988. N 4. P. 75–77. (In Russian.)

29. *Morozov N.F., Petrov Y.V. and Utkin A.A.* On limit intensity computation of dynamic pulse loadings. News Russian Science Acad. Solid Mechanics. 1988. N 5. P. 180–182. (In Russian.)

30. *Morozov N.F., Petrov Y.V. and Utkin A.A.* On cleavage analysis under approaches of fracture structural mechanics. Reports Science Acad. USSR **313** (2), 276–279 (1990). (In Russian.)

31. *Morozov N.F., Petrov Y.V. and Utkin A.A.* On Structure–temporal approach to dynamic brittle formation fracture analysis. Proceedings Leningrad Mining Institute **125** (Rock Fracture), 76–86 (1991). (In Russian.)

32. *Nazarov S.A. and Paukshto M.V.* Discrete models and Averaging in Problems of Elasticity Theory (Leningrad University Press, Leningrad, 1984, 93 pp.). (In Russian.)

33. *Neuber H.* Concentration of Stresses (Nauka, Moscow–Leningrad, 1947, 204 pp.). (In Russian.)

34. *Nikiforovsky V.S.* On solid brittle fracture kinetic character. Appl. Mech. and Tech. Phys. **18** (5), 150–157 (1976). (In Russian.)

35. *Nikiforovskii V.S. and Shemyakin E.I.* Dynamic Fracture of Solids. (Novosibirsk University Press, Novosibirsc, 1979, 271 pp.). (In Russian.)

36. *Nikolaevskii V.N.* Dynamic strength and fracture velocity. Impact, Explosion and Fracture (Nauka, Moscow, 1981, pp. 166–203). (In Russian.)

37. *Nobl B.* Wiener–Hopf Method Application for Partial Differential Equations (Nauka, Moscow, 1962, 279 pp.). (In Russian.)

38. *Novozhilov V.V.* On necessary and sufficient criterion of brittle strength. Appl. Math. and Mech. **33** (2), 212–222 (1969). (In Russian.)

39. *Novozhilov V.V.* On fundamentals of equilibrium cracks theory in elastic bodies. Appl. Math. and Mech. **33** (5), 797–802 (1969). (In Russian.)

40. *Panasjuk V.V.* Limit Equilibrium of Brittle Bodies with Cracks (Naukova Dumka, Kiev, 1968, 246 pp.). (In Russian.)

41. *Panovko J.G.* Mechanics of Deformable Solids (Nauka, Moscow, 1985, 287 pp.). (In Russian.)

42. *Parton V.Z. and Boriskovsky V.G.* Dynamic Fracture Mechanics (Nauka, Moscow, 1985, 264 pp.). (In Russian.)

43. *Parton V.Z. and Boriskovskii V.G.* Brittle Fracture Dynamics (Nauka, Moscow, 1988, 239 pp.). (In Russian.)

44. *Parton V.Z. and Morozov E.M.* Elastic–Plastic Fracture Mechanics (Nauka, Moscow, 1985, 504 pp.). (In Russian.)

45. *Petrov Y.V.* On 'quantum' nature of brittle media dynamic fracture. Reports Science Acad. USSR **321** (1), 66–68 (1991). (In Russian.)

46. *Petrov Y.V.* High-Rate Fracture of Brittle Media: Self-Abstract of Mathematical Doctoral Thesis, St. Petersburg, 1995, 25 pp. (In Russian.)

47. *Petrov Y.V.* Quantum analogy in solids fracture mechanics. Physics Solids **38** (11), 3386–3393 (1996). (In Russian.)

48. *Petrov Y.V., Smirnova S.I., Utkin A.A. and Fedorovsky G.D.* On stress state asymptotics near crack tip in thin plate. Vestnik Leningrad University, Series 1, Issue 3, 123–125 (1991). (In Russian.)

49. *Petrov Y.V. and Utkin A.A.* On dynamic strength dependence on loading rate. Physio–Chemical Material Mechanics. 1989. N 2. P. 38–42. (In Russian.)

50. *Petrov Y.V. and Utkin A.A.* On structure–temporal criterion of brittle media dynamic fracture. Vestnik Leningrad University, Series 1, Issue 4, 52–58 (1990). (In Russian.)

51. *Polezhaev Y.V.* Thermogasdynamic Aircraft Treating (Nauka, Moscow, 1986, 69 pp.). (In Russian.)

52. *Poruchikov V.B.* Methods of Dynamic Elasticity Theory (Nauka, Moscow, 1986, 328 pp.). (In Russian.)

53. *Pugachev G.S.* Fracture of Solids under Pulse Loading: Self-Abstract of Mathematical Doctoral Thesis, Leningrad, 1985, 37 pp. (In Russian.)

54. *Slepjan L.I.* Approximate model of crack dynamics. Continuum Dynamics. 1974. N 19–20. P. 101–110. (In Russian.)

55. *Slepjan L.I.* Crack Mechanics (Leningrad University Press, Leningrad, 1981, 295 pp.). (In Russian.)

56. *Stepanov G.V.* Elastic–Plastic Deformation and Material Fracture Under Pulse Loading (Naukova Dumka, Kiev, 1991, 288 pp.). (In Russian.)

57. *Urbanovich L.I., Kramchenkov E.M. and Chunosov Y.N.* Experimental study technics of low-temperature erosion, appearing under hard particles impact action. Friction and Deterioration **11** (5), 889–894 (1990). (In Russian.)

58. *Friedman J.B.* Mechanical Properties of Metal (Nauka, Moscow, 1974). In two parts. Part 1. Deformation and Fracture, 472 pp. Part 2. Mechanical Tests. Construction Strength, 367 pp. (In Russian.)

59. *Cherepanov G.P.* Elastic waves diffraction at linear cuts. Continuum Mechanics and Related Analysis Problems (Nauka, Moscow, 1972, pp. 615–622.). (In Russian.)

60. *Cherepanov G.P.* Brittle Fracture Mechanics. (Nauka, Moscow, 1974, 640 pp.). (In Russian.)

61. *Shock R.* Formation behavior under strong stresses. Mechanics Impact, Explosion, Fracture. 1981. N 26. P. 166–203. (In Russian.)

62. *Achenbach J.D.* Dynamic effects in brittle fracture. Mechanics Today (S. Neamat-Nasser, Pergamon Press, 1972. Vol. 1, pp. 1–57).

63. *Broberg K.B.* Some aspects of the mechanism of scabbing. Stress Wave Propagation in Materials (Interscience, New York, London, 1960, pp. 229–246).

64. *Campbell J.D.* The dynamic yielding of mild steel. Acta Metallurgica **1** (6), 64–80 (1953).

65. *Clark A.B. and Sanford R.J.* Static and dynamic calibration of photoelastic model materials. Proc. Soc. Exp. Stress Anal. **14** (1), 195–204 (1956).

66. *Dally J.W.* Dynamic photoelastic studies of fracture. Exp. Mech. **19**, 349–361 (1979).

67. *Dally J.W. and Barker D.B.* Dynamic measurements of initiation toughness at high loading rates. Exp. Mech. **Sept.**, 298–303 (1988).

68. *Dally J.W. and Shukla A.* Dynamic crack behaviour at initiation. Mech. Res. Comm. **6** (4), 239–244 (1979).

69. *Freund L.B.* The analysis of elastodynamic crack tip stress fields. Mechanics Today (S. Neamat-Nasser, Pergamon Press, 1972, Vol. 3. pp. 55–91).

70. *Freund L.B.* The stress intensity factor due to normal impact loading of the faces of a crack. Internat. J. Engrg. Sol. **12** (2), 179–189 (1974).

71. *Freund L.B.* Dynamic crack propagation. Mechanics Fracture (F. Erdogan, ASME AMD, 1976, Vol. 19, pp. 105–134).

72. *Griffith A.* The phenomena of rupture and flow in solids. Phil. Trans. R. Soc. London Ser. A **221**, 163–198 (1921).

73. *Homma H., Shockey D.A. and Murayama Y.* Response of cracks in structural materials to short pulse loads. J. Mech. Phys. Sol. **31** (3), 261–279 (1983).

74. *Irwin G.* Analysis of stresses and strains near the end of a crack traversing a plate. J. Appl. Mech. **24** (3), 361–364 (1957).

75. *Kalthoff J.F.* Fracture behavior under high rates of loading. Engrg. Fracture Mech. **23**, 289–298 (1986).

76. *Kalthoff J.F.* Shadow optical analysis of dynamic shear fracture. Internat. Conf. Photomechanics Speckle Metrology. SPIE–Intern. Society for Optical Engineering, Paris, **814**, 531–538 (1987).

77. *Kalthoff J.F.* Transition in the failure behavior of dynamically shear loaded cracks. Appl. Mech. Rev. **43** (5(2)), 247–250 (1977).

78. *Kalthoff J.F. and Shockey D.A.* Instability of cracks under impulse loads. J. Appl. Phys. **48**, 986–993 (1977).

79. *Kalthoff J.F., Beinert J. and Winkler S.* Measurements of dynamic stress intensity factors for fast running and arresting cracks in double-cantilever-beam specimens. Fast Fracture and Crack Arrest. ASTM STP **627**, 161–176 (1977).

80. *Kalthoff J.F., Beinert J., Winkler S. and Klemm W.* Experimental analysis of dynamic effects in different crack arrest test specimens. Crack Arrest Methodology and Applications. ASTM STP **711**, 109–127 (1980).

81. *Kalthoff J.F. and Winkler S.* Failure mode transition at high rates of shear loading. Impact'87, Internat. Conf. Impact Loading and Dynamic Behavior Materials, Bremen, May 1987, pp. 161–176.

82. *Knauss W.G.* Fundamental problems in dynamic fracture. Advances of Fracture Research. Proc. of the 6th Int. Conf. of Fracture, Delhi. (S. R. Vallury, Oxford, New York, 1984, Vol. 1, pp. 625–652).

83. *Knauss W.G. and Ravi-Chandar K.* Some basic problems in stress wave dominated fracture. Internat. J. Fracture **27**, 127–143 (1985).

84. *Kobayashi T. and Dally J.W.* Relation between crack velocity and stress intensity factor in birefringent polymers. Fast Fracture and Crack Arrest. ASTM STP **627**, 257–273 (1977). (Translated in Russian: In Mechanics of Fracture. Fast Fracture and Crack Arrest. Book of Papers. Publishing House Mir, Moscow, P.101-119. 1981).

85. *Kobayashi T., Wade B.G. and Bradley W.G.* Fracture dynamics of Homalite-100. In: Deformation and Fracture High Polymers, by H. H. Kaush, J. A. Haseele and R. I. Jaffee. (Plenum Press, New York, 1973, pp. 487–500.)

86. *Krishnaswamy S., Rosakis A.J. and Ravichandran G.* On the domain of dominance of asymptotic crack tip fields in the dynamic fracture of metals: an investigation based on bifocal caustics and three-dimensional dynamic numerical simulations. SM Report 88–18, Pasadena, 1988, 27 pp.

87. *Lawn B.R. and Wilshaw T.R.* Indentation fracture: principles and application. J. Mater. Sci. **10** (6), 1049–1081 (1975).

88. *Lee Y.J. and Freund L.B.* Fracture initiation due to asymmetric impact loading of an edge cracked plate. ASME J. Appl. Mech. **57**, 104–111 (1990).

89. *Ma C.C. and Freund L.B.* The extent of the stress intensity factor during crack growth under dynamic loading conditions. ASME J. Appl. Mech. **53**, 303–310 (1986).

90. *Manogg P.* Schattenoptische Messung der spezifischen Bruchenergie wahrend des Bruchvorgangs bei Plexiglas. Proc. of the Internat. Conf. on Physics of Non-Crystalline Solids, Delft, 1964, pp.481–490.

91. Honeycombe R.W.K. Steels, Microstructure and Properties (Edward Arnold, Cambridge, 1980, 270 pp.)

92. *Morozov N.F. and Paukshto M.V.* On the crack simulation and solution in the lattice. ASME J. Appl. Mech. **58**, 290–292 (1991).

93. *Morozov N.F. and Petrov Y.V.* New principles of testing of dynamic strength of materials. Macro- and Micromechanical Aspects of Fracture. Proc. EUROMECH-291, St. Petersburg, 22–27 June 1992, p. 25.

94. *Morozov N.F. and Petrov Y.V.* On the macroscopic parameters of brittle fracture . Arch. Mech. Internat. J. **48** (5), 825–833 (1996).

95. *Morozov N.F. and Petrov Y.V.* The problems of high rate loading: the new criterion of fracture, the erosion, the asymmetric impact loading. IUTAM Constitutive Relation in High/Very High Strain Rates (Springer-Verlag, Tokyo, 1996, pp. 225–232).

96. *Morozov N.F., Petrov Y.V. and Utkin A.A.* New explanation of some effects of brittle fracture by impact loading. Advances of Fracture Research. Proc. of the 7th Int. Conf. of Fracture, Houston. (S. Neamat-Nasser, Oxford, 1989. Vol. 6, pp. 3703–3711.

97. *Morozov N.F., Urbanovich L.I., Kramchenkov E.M. and Petrov Y.V.* Influence of mechanical and physical properties of both solid particles and target materials on the critical impact velocity. Proc. Internat. Conf. Arzamas-16, Nizhny Novgorod, 1994, p. 47.

98. *Petrov Y.V.* Criterion of dynamic fracture of brittle solids. Proc. 8th Internat. Conf. Fracture, Kiev, 8–14 June 1993, p. 97.

99. *Petrov Y.V. and Morozov N.F.* On the modeling of fracture of brittle solids. ASME J. Appl. Mech. **61**, 710–712 (1994).

100. *Petrov Y.V. and Utkin A.A.* On fracture initiation due to asymmetric impact loading. Experiments and Macroscopic Theory in Crack Propagation. Proc. EUROMECH-326, Kielce, 25–28 Sept. 1994, p. 23.

101. *Ravi-Chandar K.* Crack propagation under pressure and shear. Experiments and Macroscopic Theory Crack Propagation. Proc. EUROMECH-326, Kielce, 25–28 Sept. 1994, p. 37.

102. *Ravi-Chandar K. and Knauss W.G.* An experimental investigation into dynamic fracture: 1. Crack initiation and arrest. Internat. J. Fracture **25**, 247–262 (1984).

103. *Ravi-Chandar K. and Knauss W.G.* An experimental investigation into dynamic fracture: 2. Microstructural aspects. Internat. J. Fracture **26**, 65–80 (1984).

104. *Ravi-Chandar K. and Knauss W.G.* An experimental investigation into dynamic fracture: 3. On steady state crack propagation and crack branching. Internat. J. Fracture **26**, 141–154 (1984).

105. *Ravi-Chandar K. and Knauss W.G.* On the characterization of the transient stress field near the tip of a crack. ASME J. Appl. Mech. **54**, 72–78 (1987).

106. *Rosakis A.J. and Ravi-Chandar K.* On crack tip stress state: an experimental evaluation of three dimensional effects. Internat. J. Solids and Structures **22**, 121–128 (1986).

107. *Rosakis A.J., Mason J.J. and Ravichandran G.* Full field measurements of dynamic deformation field around a growing adiabatic shear band at the crack tip of a dynamically loaded crack or notch. SM Report 93-41, Pasadena, 1993. 30 pp.

108. *Rosakis A.J., Zehnder A.T. and Narasimhan R.* Caustics by reflection and their application to elastic–plastic and dynamic fracture mechanics. Opt. Engrg. **27** (8), 596–610 (1988).

109. *Seaman L., Curran D.R., Aidun J.B. and Cooper T.* A microstatistical model for ductile fracture with rate effects. Nuclear Engrg. Design **105**, 35–42 (1987).

110. *Seaman L., Curran D.R. and Murri W.J.* A continuum model for dynamic tensile microfracture and fragmentation. J. Appl. Mech. **52**, 593–600 (1985).

111. *Shih C.F.* Small-scale yielding analysis of mixed mode plain strain crack problems. Fracture Anal. ASTM STP **560**, 187–210 (1974).

112. *Shockey D.A., Erlich D.C., Kalthoff J.F. and Homma H.* Short-pulse fracture mechanics. Engrg. Fracture Mech. **23**, 311–319 (1986).

113. *Shockey D.A., Seaman L. and Curran D.R.* Application of microstatistical fracture mechanics to dynamic fracture problems. In: Material. Behavior under High Stress and Ultrahigh Loading Rates, by J. Mescal and V. Welse (Plenum Publishing Inc., New York, 1983, pp. 273–293).

114. *Sih G.C.* Some elastodynamics problems of cracks. Internat. J. Fracture Mech. **4**, 51–68 (1968).

115. *Sih G.C. and Macdonald B.* Fracture mechanics applied to engineering problems. Strain energy density fracture criterion. Engrg. Fracture Mech. **6** (2), 361–386 (1974).

116. *Smith G.C.* An Experimental Investigation of the Dynamic Fracture of a Brittle Material. Ph. D. Thesis, Pasadena, 1975, 93 pp.

117. *Theocaris P.S. and Gdoutos E.* An optical method for determining opening-mode and edge-sliding-mode stress intensity factors. J. Appl. Mech. **39**, 91–97 (1972).

118. *Theocaris P.S. and Papadopoulos G.A.* The dynamic behavior of sharp V-notches under impact loading. Internat. J. Solids and Structures **23** (12), 1581–1600 (1987).

119. *Thomson R., Hsieh C. and Rana V.* Lattice trapping of fracture cracks. J. Appl. Phys. **42** (8), 3154–3160, (1971).

120. Treatise on Material Science and Technology, Vol. 16. Erosion. Book of papers. (Ed. by C.M. Preece. New York, San Francisco, London, Ltd., 1979, 464 pp.).

121. *Wieghardt K.* Uber das Spalten und Zerreisen elastischer Körper. Z. Math. Phys. **55** (7/8), 60–103 (1907).

122. *Klepaczko J.R.* Modeling of transition from brittle to ductile fracture with statistic approach and new local fracture criterion. Endommagement et Rupture Dynamiques, DYMAT, 6éme Journee Nationale, Metz, 6 November 1990, pp. 1–8.

123. *Panin V.E., Gridnev Y.V. and Danilov V.I.al.* Structural Levels of Deforming and Fracture (Nauka, Novosibirsk, 1990).

124. *Meyers M.A.* Dynamic Behavior of Materials (J. Wiley, New York, 1994, Chapter 13, pp. 323–381).

# Springer
# and the
# environment

At Springer we firmly believe that an
international science publisher has a
special obligation to the environment,
and our corporate policies consistently
reflect this conviction.
We also expect our business partners –
paper mills, printers, packaging
manufacturers, etc. – to commit
themselves to using materials and
production processes that do not harm
the environment. The paper in this
book is made from low- or no-chlorine
pulp and is acid free, in conformance
with international standards for paper
permanency.

 Springer

Printing: Mercedes-Druck, Berlin
Binding: Buchbinderei Lüderitz & Bauer, Berlin